AIRCRAFT CALCULATIONS

A new edition of

AIRCRAFT MATHEMATICS

by

S. A. WALLING
Senior Master R.N. (Ret.)

and

J. C. HILL, B.A. (CANTAB.)
Education Department, Cambridge University Press

CAMBRIDGE
AT THE UNIVERSITY PRESS
1942

This edition of *Aircraft Mathematics* has been planned in collaboration with Air Commodore J. A. Chamier, C.B., C.M.G., O.B.E., D.S.O. and the headquarters staff of the Air Training Corps, and is published with their approval, in its revised form as

AIRCRAFT CALCULATIONS

Aircraft Mathematics

First edition	*August* 1941
Reprinted (with corrections)	*October* 1941
Second edition	*January* 1942
Canadian edition	*November* 1941
(The Macmillan Company, Toronto)	
American edition	*June* 1942
(The Macmillan Company, New York)	

Aircraft Calculations

First edition	*October* 1942

CAMBRIDGE
UNIVERSITY PRESS

University Printing House, Cambridge CB2 8BS, United Kingdom

Cambridge University Press is part of the University of Cambridge.

It furthers the University's mission by disseminating knowledge in the pursuit of education, learning and research at the highest international levels of excellence.

www.cambridge.org
Information on this title: www.cambridge.org/9781316619858

First paperback edition 2016

A catalogue record for this publication is available from the British Library

ISBN 978-1-316-61985-8 Paperback

PREFACE

In consideration of the modifications made recently to the syllabus of mathematics for Air Training Corps cadets and in consultation with the Administrative Staff of A.T.C. headquarters, it has been decided to remodel this book so that the essential material shall be more readily accessible to the student.

The book has now been arranged in two parts and its title changed.

Part I contains the revisionary work and exercises which are definitely within the scope of the syllabus.

Part II is made up of exercises and problems involving the simple fundamental rules of algebra, geometry, logarithms, and trigonometry.

All the work, as far as possible, in both Parts I and II is based upon Service conditions or flying problems so that the cadet shall be able to see the value of a training in simple mathematics by its practical application.

Although knowledge of the principles and ability to work the exercises in Part II are not essential for the cadet in order to obtain the necessary proficiency, they have been included for the benefit of those many cadets who will find instruction and possible enjoyment in solving the varied and typical Service problems.

A knowledge of logarithms, for example, and the ability to use logarithm tables, although not essential to a pilot-observer, will, without doubt, enable him to use with intelligence and understanding the various logarithmic computors that are widely employed in practice. The Tables of Logarithms printed on pages 132–143 are included by the kind permission of the Cambridge Local Examinations Syndicate.

With few exceptions, which consist mostly of additional exercises and the omission of problems based exclusively upon navigation, the content of the book as it now stands is the same as that in previous editions of *Aircraft Mathematics*, except that it appears in a different sequence.

For this reason little difficulty should be experienced in using this book with classes where it has already been adopted in its original form.

The authors would like to express their appreciation of the very helpful criticisms and suggestions that have been made to them by Air Commodore J. A. Chamier, C.B., C.M.G., O.B.E., D.S.O.

S. A. W.
J. C. H.

July 1942

For those students requiring an up-to-date text book of navigation, both theory and practice, with a comprehensive study of meteorology and a course of practical work including exercises on the plotting chart and logarithmic computor (both of which are supplied with the book) a companion book to this is published by Cambridge University Press, entitled

AIRCRAFT NAVIGATION

Students requiring further exercises and problems are recommended to write for particulars of a

SERIES OF 100 TEST CARDS FOR REVISION

of previous work done in

AIRCRAFT CALCULATIONS

CONTENTS

PART ONE

ELEMENTARY ARITHMETIC

SIGNS AND SYMBOLS

GRAPHS

GEOMETRY

PART TWO

PART ONE

ELEMENTARY ARITHMETIC

FACTORS AND LEAST COMMON MULTIPLE

Factors.

When a number divides exactly into another number, the first is said to be a *factor* of the second.

EXAMPLE. 3 is a factor of 15. So is 5 a factor of 15. Therefore the factors of 15 are 3 and 5.

Many numbers have more than two factors, such as

$$24=2\times 12=2\times 2\times 6=2\times 2\times 2\times 3.$$

Some numbers have no factors, other than the number itself and 1.

Thus the only factors of 7 are 7 and 1. Such numbers are called *prime numbers.*

So that it follows that any number expressed in factors can be expressed in prime numbers.

EXAMPLE. $182=2\times 91=2\times 7\times 13$, each of the final factors being a prime number.

These factors are called *prime factors.*

EXERCISE I.

1. Which of the following are prime numbers: 1, 3, 7, 9, 11, 15, 17, 51, 53, 71, 73?

2. Find the prime factors of:

(*a*) 16. (*b*) 32. (*c*) 71. (*d*) 85.
(*e*) 168. (*f*) 211. (*g*) 251. (*h*) 484.
(*i*) 571. (*j*) 9261. (*k*) 11025. (*l*) 1111.

Least Common Multiple (L.C.M.).

A number is said to be a *multiple* of each of its factors.

Thus 180, 120 and 60 are each multiples of 20. They are also multiples of 12 and are consequently *common multiples* of 20 and 12.

The *smallest* number which is a common multiple of any factors is called the *least common multiple* (L.C.M.).

EXERCISE II.

1. Find the L.C.M. of:

(*a*) 4 and 6. (*b*) 9 and 18. (*c*) 18 and 27.
(*d*) 7 and 21. (*e*) 24 and 36. (*f*) 3, 4 and 5.
(*g*) 6, 8 and 12. (*h*) 2, 12, 6 and 18. (*i*) 3, 6, 15 and 20.
(*j*) 3, 4, 12 and 18. (*k*) 5, 10, 15 and 20. (*l*) 1, 2, 3, 4 and 5.

FRACTIONS

A fraction is always a part of something. For example: three farthings are written as $\frac{3}{4}d.$ to denote that they are three quarters of a penny.

In all fractions the number above the line is called the *numerator*, and that below the line is called the *denominator*.

All fractions should be expressed in their *lowest terms*. This can be done by dividing the numerator and denominator by any common factors:

$$\frac{\overset{3}{\cancel{9}}}{\underset{4}{\cancel{12}}}=\frac{3}{4}.$$

All fractions that have the numerator smaller than the denominator are termed *proper fractions*.

Occasionally you will meet with fractions having the numerator larger than the denominator, e.g. $\frac{21}{9}$.

Such fractions are called *improper fractions* and should never be left as such in your answer. They should be changed to a *mixed number* and reduced to their lowest terms. Thus

$$\frac{21}{9}=2\frac{3}{9}=2\frac{1}{3}.$$

It is often necessary when working with fractions to change mixed numbers into improper fractions, but the change back must be made in the final answer.

EXERCISE III.

1. Reduce the following fractions to their lowest terms:

(*a*) $\frac{2}{6}$. (*b*) $\frac{3}{6}$. (*c*) $\frac{6}{8}$. (*d*) $\frac{12}{16}$. (*e*) $\frac{10}{15}$. (*f*) $\frac{10}{25}$.
(*g*) $\frac{18}{30}$. (*h*) $\frac{25}{28}$. (*i*) $\frac{27}{81}$. (*j*) $\frac{64}{84}$. (*k*) $\frac{28}{49}$. (*l*) $\frac{45}{100}$.
(*m*) $\frac{242}{1111}$. (*n*) $\frac{370}{555}$. (*o*) $\frac{216}{243}$. (*p*) $\frac{220}{924}$. (*q*) $\frac{572}{1012}$. (*r*) $\frac{1617}{1815}$.
(*s*) $\frac{1008}{1728}$. (*t*) $\frac{2940}{4620}$. (*u*) $\frac{264}{1760}$. (*v*) $\frac{39}{169}$. (*w*) $\frac{465}{705}$. (*x*) $\frac{729}{945}$.

2. Express the following mixed numbers as improper fractions:

(*a*) $1\frac{3}{4}$. (*b*) $3\frac{1}{2}$. (*c*) $4\frac{3}{4}$. (*d*) $8\frac{1}{7}$. (*e*) $6\frac{4}{5}$. (*f*) $4\frac{3}{100}$.
(*g*) $4\frac{57}{100}$. (*h*) $5\frac{7}{22}$. (*i*) $20\frac{6}{13}$. (*j*) $41\frac{3}{11}$. (*k*) $13\frac{2}{15}$. (*l*) $17\frac{7}{31}$.
(*m*) $41\frac{11}{32}$. (*n*) $25\frac{2}{25}$. (*o*) $10\frac{17}{18}$. (*p*) 15.

3. Change these improper fractions into mixed numbers:

(*a*) $\frac{7}{4}$. (*b*) $\frac{8}{3}$. (*c*) $\frac{18}{5}$. (*d*) $\frac{31}{2}$. (*e*) $\frac{17}{12}$. (*f*) $\frac{22}{7}$.
(*g*) $\frac{507}{100}$. (*h*) $\frac{181}{17}$. (*i*) $\frac{630}{23}$. (*j*) $\frac{1143}{112}$. (*k*) $\frac{3764}{97}$. (*l*) $\frac{1141}{29}$.
(*m*) $\frac{9250}{157}$. (*n*) $\frac{4439}{47}$. (*o*) $\frac{1790}{101}$. (*p*) $\frac{7441}{89}$.

4. Fill in the gaps:

(*a*) $\frac{5}{8}=\frac{}{24}$. (*b*) $\frac{2}{3}=\frac{}{36}$. (*c*) $\frac{5}{9}=\frac{55}{}$.
(*d*) $\frac{6}{17}=\frac{18}{}$. (*e*) $\frac{3}{7}=\frac{}{21}=\frac{}{49}$. (*f*) $\frac{9}{10}=\frac{}{20}=\frac{}{100}$.
(*g*) $7=\frac{}{3}=\frac{63}{}$. (*h*) $13=\frac{}{2}=\frac{}{7}=\frac{117}{}$.

Addition and Subtraction of Fractions.

When adding or subtracting fractions, first bring all fractions to a common denominator (L.C.M.), find the sum and/or difference of the numerators and reduce the answer, if necessary, to its lowest terms.

EXAMPLE. $\frac{1}{8}+\frac{1}{4}-\frac{1}{6}=\frac{3+6-4}{24}=\frac{5}{24}$.

When adding or subtracting mixed numbers *do not make improper fractions*. Merely find the total of the whole numbers and add this to the fractional answer, reduced to lowest terms.

EXAMPLE.
$$3\tfrac{1}{4}+3\tfrac{5}{6}-1\tfrac{3}{8}$$
$$=5+\tfrac{1}{4}+\tfrac{5}{6}-\tfrac{3}{8}$$
$$=5+\frac{6+20-9}{24}$$
$$=5+\tfrac{17}{24}=5\tfrac{17}{24}.$$

EXERCISE IV.

1. Find the value of:

(a) $\frac{1}{2}+\frac{1}{4}$. (b) $\frac{2}{5}+\frac{3}{10}$. (c) $\frac{3}{14}+\frac{5}{21}$. (d) $\frac{4}{15}+\frac{3}{10}$.
(e) $\frac{5}{12}+\frac{7}{15}$. (f) $\frac{4}{11}+\frac{2}{33}+\frac{5}{22}$. (g) $2\frac{1}{2}+3\frac{1}{3}$. (h) $\frac{5}{6}+1\frac{5}{12}$.
(i) $4\frac{3}{8}+3\frac{5}{12}$. (j) $1\frac{2}{5}+1\frac{1}{4}+\frac{7}{10}$.
(k) $\frac{8}{9}+3\frac{5}{6}+1\frac{11}{12}$. (l) $2\frac{3}{10}+1\frac{61}{100}+3\frac{7}{1000}$.

2. Find the value of:

(a) $\frac{3}{4}-\frac{1}{4}$. (b) $\frac{1}{3}-\frac{1}{7}$. (c) $\frac{3}{4}-\frac{1}{6}$. (d) $\frac{7}{15}-\frac{3}{10}$.
(e) $\frac{7}{12}-\frac{7}{18}$. (f) $3\frac{5}{6}-2\frac{1}{3}$. (g) $3\frac{7}{8}-1\frac{3}{4}$. (h) $3\frac{5}{8}-1\frac{3}{4}$.
(i) $4\frac{3}{8}-2\frac{7}{12}$. (j) $7\frac{1}{2}-3\frac{3}{5}$. (k) $6\frac{4}{7}-5\frac{3}{5}$. (l) $5\frac{3}{7}-4\frac{1}{2}$.

3. Simplify:

(a) $\frac{1}{4}+\frac{1}{2}-\frac{2}{5}$. (b) $1\frac{1}{10}-1\frac{1}{100}+1\frac{1}{1000}$. (c) $4\frac{1}{2}+2\frac{1}{6}-2\frac{1}{3}$.
(d) $4\frac{1}{8}-2\frac{1}{4}+4\frac{3}{4}$. (e) $2\frac{1}{6}-1\frac{5}{6}+3\frac{1}{3}$. (f) $8\frac{7}{12}-2\frac{17}{18}+3\frac{1}{6}$.
(g) $3\frac{7}{15}-1\frac{13}{20}+\frac{17}{25}$. (h) $4\frac{3}{8}-4\frac{1}{2}-\frac{5}{8}+10$. (i) $6\frac{1}{2}-4\frac{1}{6}+3\frac{1}{12}-3\frac{3}{4}$.
(j) $11\frac{11}{21}+\frac{1}{14}-2\frac{12}{49}-2\frac{1}{7}$. (k) $6\frac{3}{22}-2\frac{17}{33}-1\frac{1}{2}$.
(l) $3\frac{5}{9}-2\frac{1}{3}+1\frac{5}{12}+1\frac{5}{6}$.

4. Take the smaller from the greater of:

(a) $\frac{5}{6}$ and $\frac{4}{7}$. (b) $\frac{77}{1000}$ and $\frac{8}{100}$. (c) $\frac{7}{8}$ and $\frac{7}{9}$.
(d) $7\frac{5}{6}$ and $7\frac{7}{8}$. (e) $\frac{8}{21}$ and $\frac{11}{28}$. (f) $\frac{15}{26}$ and $\frac{29}{39}$.

5. Arrange in order of magnitude, placing greatest first:

$$\tfrac{2}{3};\ \tfrac{5}{6};\ \tfrac{3}{4};\ \tfrac{4}{5};\ \tfrac{7}{10}.$$

6. What would bolts of the following diameters measure if tested with a vernier micrometer reading $\frac{1}{1000}$ths of an in.?

(a) $\frac{3}{8}$ in. (b) $\frac{1}{4}$ in. (c) $\frac{7}{8}$ in. (d) $\frac{11}{16}$ in.

PROBLEMS

Exercise V.

1. I fly $\frac{7}{8}$ of my journey and motor the remaining 25 miles. What is the total length of my journey?

2. A plane has to fly 630 miles from an aerodrome to Berlin. How much farther has it to go after reaching the Dutch coast, $\frac{2}{7}$ of its total journey?

3. The resistance R which a conductor offers to the passage of electric current is measured in *ohms*.

If several wires are joined together in series—that is, end to end so as to form one long wire—the total resistance of such a wire is the sum of the separate resistances.

Example. What would be the total resistance of the following three wires in series?

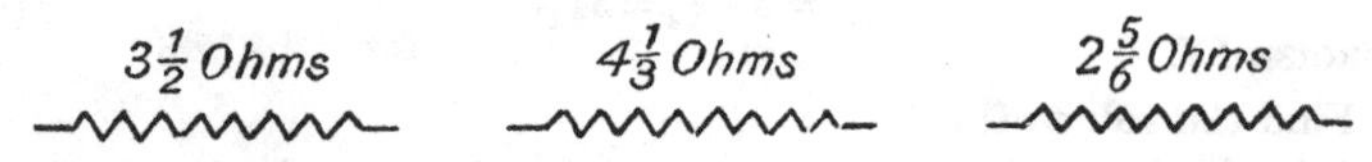

When in series we have:

$3\frac{1}{2}$ $4\frac{1}{3}$ $2\frac{5}{6}$

$$\begin{aligned}\text{The total resistance } R &= 3\tfrac{1}{2}+4\tfrac{1}{3}+2\tfrac{5}{6}\\ &= 9\,\frac{3+2+5}{6}\\ &= 9\tfrac{10}{6}\\ &= 10\tfrac{2}{3}\text{ ohms.}\end{aligned}$$

Find the total resistances of the following wires when joined in series:

(*a*) $2\frac{2}{3}$ ohms and $5\frac{3}{4}$ ohms.

(*b*) $1\frac{3}{5}$ ohms, $3\frac{1}{3}$ ohms and $20\frac{5}{6}$ ohms.

(*c*) $4\frac{1}{2}$ ohms, $3\frac{1}{7}$ ohms and $5\frac{5}{9}$ ohms.

4. When two or more wires are joined in parallel—or side by side—they afford alternative paths for the current. Thus

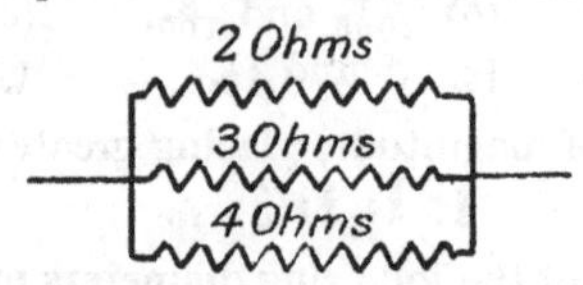

The total resistance is less than that of the smallest branch.

It is found in this way. The wire of resistance 2 ohms is said to have a conductance of $\frac{1}{2}$, i.e. the conductance is obtained by inverting the resistance.

Thus a resistance r has a conductance of $\frac{1}{r}$, where r may be any value.

To find the total resistance of a circuit in parallel, find the total of the separate conductances and invert this to find the total resistance.

In the above example, if R is the total resistance, then

$$\frac{1}{R}=\frac{1}{2}+\frac{1}{3}+\frac{1}{4}=\frac{6+4+3}{12}=\frac{13}{12}.$$

$$\therefore\ R=\tfrac{12}{13}\text{ ohms.}$$

Find the resistances of the following wires in parallel:

(*a*) $1\frac{1}{2}$ ohms and $2\frac{1}{2}$ ohms. (*b*) 2, 4 and 8 ohms.

(*c*) $1\frac{1}{4}$, $2\frac{1}{2}$ and $3\frac{3}{4}$ ohms. (*d*) $1\frac{7}{8}$, 5 and $1\frac{2}{3}$ ohms.

5. A condenser is used in wireless telegraphy to store electrical energy. In its simplest form it consists of two parallel, oblong metal plates separated by a non-conducting substance called a "di-electric". Air, glass, paraffin wax and certain oils are all used as dielectrics.

In drawings of circuits a condenser is shown thus: where A and B are the plates viewed edgewise and the dielectric is in the space between.

Large condenser capacities are measured in *Farads* and the capacity is a measure of the storing power.

Capacities of small condensers are expressed in microfarads, where

1 Farad $=1{,}000{,}000$ microfarads

or 1 microfarad (mf.) $=\dfrac{1}{\text{millionth}}$ of a Farad.

Condensers act oppositely to resistances.

When in *parallel* the total capacity is the sum of the separate capacities.

In the above circuit the total capacity

$=5+3+4=12$ microfarads.

When in *series* they are treated like resistances in parallel and the total capacity is found by adding the reciprocals and inverting.

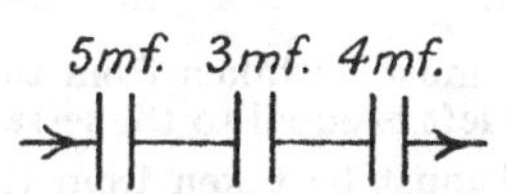

Thus if S is the total capacity,

$$\frac{1}{S}=\frac{1}{5}+\frac{1}{3}+\frac{1}{4}=\frac{12+20+15}{60}=\frac{47}{60}.$$

$$\therefore\ S=\tfrac{60}{47}=1\tfrac{13}{47}\text{ microfarads.}$$

Find the total capacities in microfarads of the following circuits:

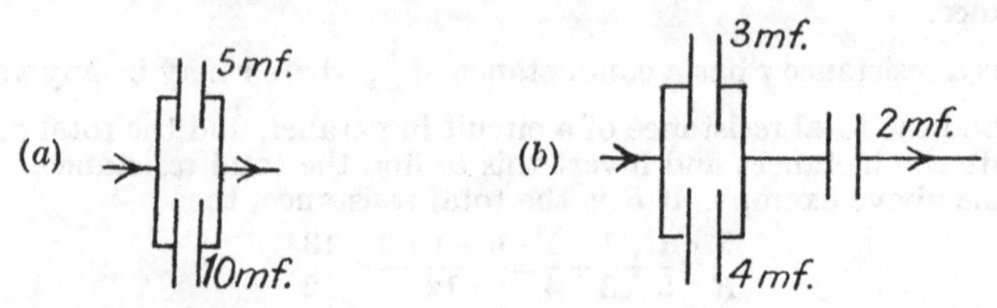

(NOTE. The 3 mf. and 4 mf. in parallel act as one single condenser of 7 mf. capacity. Thus we have 7 mf. and 2 mf. in series.)

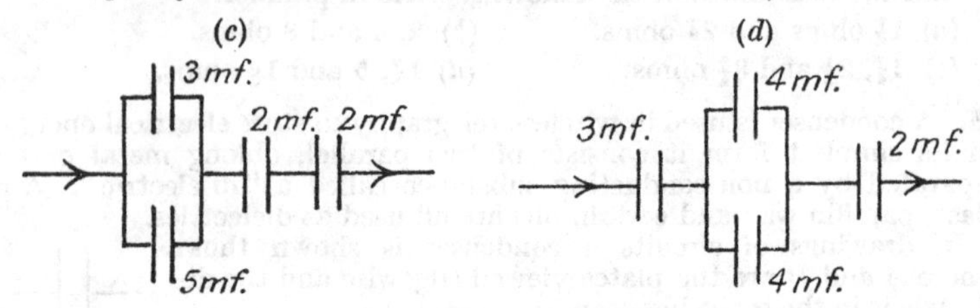

6. (*a*) If the total resistance of these wires is $12\frac{1}{8}$ ohms, what is the resistance of the part x?

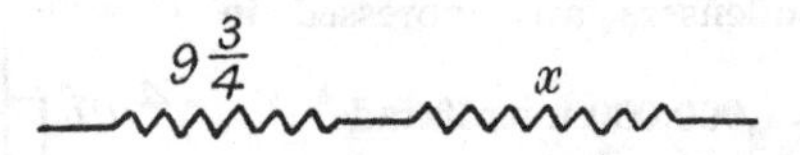

(*b*) If the total capacity of these three condensers is $7\frac{1}{12}$ microfarads, what is the capacity of condenser x?

7. A pilot used $\frac{1}{4}$ of his petrol supply flying North. Then $\frac{1}{5}$ of his total supply flying East. What fraction of his tank capacity remained? If he was left with 99 gallons, what was his original supply in gallons?

8. Two petrol tanks are of equal capacity. From the first $\frac{2}{3}$ has been used and from the second $\frac{2}{5}$.

(*a*) How much petrol must be taken from the tank with the larger quantity so that what is left is equal to the lesser?

(*b*) How much petrol must be taken from the larger quantity and transferred to the lesser, to equalise the amounts in each tank?

Answer, in each case, as fraction of whole tank.

Multiplication of Fractions.

Always change mixed numbers to improper fractions before multiplying. Thus

$$2\tfrac{2}{3} \times 5\tfrac{1}{4} = \frac{\overset{2}{\cancel{8}}}{\cancel{3}} \times \frac{\overset{7}{\cancel{21}}}{\cancel{4}} = 14.$$

EXERCISE VI.

1. Find the value of:

(a) $\frac{3}{5} \times \frac{4}{7}$. (b) $\frac{3}{8}$ of $\frac{3}{5}$. (c) $\frac{5}{16}$ of 12. (d) $21 \times \frac{3}{7}$.
(e) $\frac{15}{16} \times \frac{20}{21}$. (f) $\frac{33}{40}$ of $\frac{72}{121}$. (g) $\frac{63}{77} \times \frac{22}{35}$. (h) $\frac{51}{60}$ of $\frac{5}{68}$.

2. Find the value of:

(a) $1\frac{1}{4} \times \frac{2}{3}$. (b) $\frac{3}{5}$ of $2\frac{1}{2}$. (c) $2\frac{1}{4} \times 1\frac{1}{3}$. (d) $3\frac{3}{8}$ of $3\frac{1}{3}$.
(e) $5\frac{1}{3} \times 1\frac{1}{4}$. (f) $11\frac{4}{7} \times 5\frac{4}{9}$. (g) $1\frac{11}{25}$ of $1\frac{19}{36}$. (h) $2\frac{5}{12} \times 1\frac{7}{29}$.

3. Simplify:

(a) $\frac{9}{10} \times \frac{5}{8} \times \frac{4}{9}$. (b) $5\frac{1}{4} \times \frac{5}{7} \times 2\frac{2}{5}$. (c) $\frac{7}{12}$ of $9\frac{1}{3} \times 3\frac{3}{5}$.
(d) $2\frac{13}{16} \times 1\frac{1}{3} \times 1\frac{5}{9}$. (e) $2\frac{1}{10} \times \frac{5}{7}$ of 24. (f) $3\frac{3}{4} \times 1\frac{7}{8} \times 4\frac{4}{5}$.

4. Find the distances flown in the following cases:

	Speed	Flying hours	Distance (to nearest mile)
(a)	110 m.p.h.	$2\frac{1}{2}$	
(b)	$96\frac{1}{2}$,,	$1\frac{3}{4}$	
(c)	215 ,,	$2\frac{1}{3}$	
(d)	$172\frac{1}{2}$,,	$1\frac{3}{8}$	
(e)	$223\frac{1}{2}$,,	$1\frac{2}{3}$	

PROBLEMS

EXERCISE VII.

1. A watch gains $1\frac{2}{5}$ sec. per day. What will it gain in the following times (to nearest $\frac{1}{5}$ sec.)?

(a) $3\frac{1}{2}$ days. (b) 7 days. (c) $2\frac{3}{4}$ days.

2. A plane is climbing steadily at 142 m.p.h. If the plane's altitude changes at $\frac{2}{15}$ of this speed, at how many miles per hour is the altitude changing?

3. If petrol weighs $7\frac{1}{5}$ lb. per gallon and a plane flying at a certain speed uses 30 gal. per hour, by how much is the weight of the plane reduced after $3\frac{1}{4}$ hours flight at this speed?

4. $A = L \times B$.

This is a short way of expressing

$$\text{Area} = \text{Length} \times \text{Breadth}$$

for a rectangular (oblong) plate.

The length and breadth must *both* be in the same units; for example, (a) both in feet, or (b) both in inches, or (c) both in centimetres.

The area is then expressed in (*a*) *square* feet, (*b*) *square* inches, or (*c*) *square* centimetres.

Find the areas of the following oblong plates and name the units of area in each case:

	L	B	Area
(*a*)	$2\frac{1}{2}$ ft.	$1\frac{3}{4}$ ft.	
(*b*)	$5\frac{2}{3}$ in.	$3\frac{1}{3}$ in.	
(*c*)	$6\frac{1}{4}$ in.	$4\frac{2}{3}$ in.	
(*d*)	$1\frac{1}{4}$ ft.	$2\frac{1}{12}$ ft.	

5. The area of a circle is $\pi \times r \times r$, where π is a fixed number, as nearly as possible $\frac{22}{7}$ when represented as a fraction, and r is the radius of the circle.

It is often more convenient to measure the diameter of a circle or of a wire, or rod, so that the formula becomes

$$\text{Area} = \frac{22}{7} \times \frac{D}{2} \times \frac{D}{2},$$

where D is the diameter, i.e. $\frac{11}{14} \times D \times D$.

Find the cross-sectional area of these circularly sectioned metal struts whose diameters are:

(*a*) $\frac{7}{8}$ in. (*b*) $\frac{7}{16}$ in. (*c*) $\frac{3}{4}$ in. (*d*) $\frac{5}{16}$ in.

6. Find the cross-sectional area of a seven-strand wire rope. Each strand is, in section, a circle of $D = \frac{1}{8}$ in.

7. From these details of a Bristol Beaufort: Wingspan 57 ft. 10 in.; Length 44 ft. 2 in.; Height 14 ft. 3 in., find the wingspan, length and height of a model. Scale: $\frac{1}{4}$ in. = 1 ft.

8. The fuselage of a certain monoplane contains in its construction: 4 longerons each $38\frac{1}{3}$ ft. long, and 14 struts each $7\frac{1}{3}$ ft. long. What total length (in feet) of Balsa wood strip would be needed to make these parts in a model $\frac{3}{32}$ full size?

9. Two of the various kinds of thermometer in use for measuring temperature are the Fahrenheit and the Centigrade thermometers. On the Fahrenheit thermometer the freezing point of water is 32° F. and the boiling point of water 212° F. (a difference of 180° F.). On the Centigrade thermometer the freezing point of water is 0° C. and the boiling point 100° C. (a difference of 100° C.).

To convert from ° C. to ° F. first multiply by $\frac{180}{100}$, i.e. $\frac{9}{5}$, and add 32°.

EXAMPLE. Change 60° C. to ° F.

$$(60 \times \tfrac{9}{5}) + 32 = 108 + 32 = 140° \text{ F.}$$

To convert a temperature from ° F. to ° C. first subtract 32° and then multiply by $\frac{5}{9}$.

EXAMPLE. What temperature in ° C. is the same as 104° F.?

$$(104 - 32)\,\tfrac{5}{9} = 72 \times \tfrac{5}{9} = 40° \text{ C.}$$

Convert these ° F. into ° C.:

(*a*) 158° F. (*b*) 185° F. (*c*) 71° F. (*d*) 14° F.

Convert these ° C. into ° F.:

(*e*) 35° C. (*f*) $62\frac{1}{2}$° C. (*g*) 17° C. (*h*) −25° C.

10. In an aerial battle $\frac{1}{10}$ of the total attacking force is destroyed in the first encounter and $\frac{1}{9}$ of the remainder in a second engagement. If 16 machines return, how many started out?

Division of Fractions.

Always change mixed numbers to improper fractions. Then invert the divisor and multiply. Thus:

$$2\frac{3}{5} \div \frac{11}{15} = \frac{13}{\cancel{5}} \times \frac{\overset{3}{\cancel{15}}}{11} = \frac{39}{11} = 3\frac{6}{11}.$$

EXERCISE VIII.

1. Find the value of:

(*a*) $\frac{7}{8} \div 5$. (*b*) $\frac{3}{5} \div 4$. (*c*) $8 \div \frac{2}{3}$. (*d*) $25 \div \frac{5}{8}$.

(*e*) $\frac{9}{14} \div \frac{18}{35}$. (*f*) $\frac{3}{7} \div \frac{9}{49}$. (*g*) $\frac{25}{42} \div \frac{35}{36}$. (*h*) $1 \div \frac{17}{25}$.

2. Simplify:

(*a*) $33\frac{1}{3} \div 6\frac{1}{4}$. (*b*) $18\frac{3}{4} \div 3\frac{1}{8}$. (*c*) $23\frac{4}{5} \div 4\frac{1}{4}$.

(*d*) $8\frac{1}{10} \div 4\frac{1}{2}$. (*e*) $2\frac{1}{2} \div \frac{1}{10}$. (*f*) $21\frac{7}{11} \div \frac{17}{132}$.

3. $\text{Speed} = \dfrac{\text{Distance travelled}}{\text{Time taken}}$, i.e. Distance ÷ Time.

If the distance is in miles and the time in hours, the speed is in miles per hour. Solve the following:

	Distance	Time	Speed in m.p.h.
(*a*)	305 miles	$2\frac{1}{2}$ hours	
(*b*)	418 ,,	$2\frac{3}{4}$,,	
(*c*)	220 ,,	$1\frac{1}{5}$,,	
(*d*)	$20\frac{1}{2}$,,	$\frac{1}{12}$ hour	

4. $\text{Time taken} = \dfrac{\text{Distance travelled}}{\text{Speed}}$, i.e. Time taken = Distance ÷ Speed.

Find the flying time in hours (or fractions of an hour) in the following:

(*a*) 210 miles flown at 120 m.p.h.

(*b*) $287\frac{1}{2}$,, ,, 115 ,,

(*c*) 399 ,, ,, 126 ,,

(*d*) 738 ,, ,, 180 ,,

PROBLEMS

EXERCISE IX.

1. A photographic plate has an area of $13\frac{13}{16}$ sq. in. Its length is $4\frac{1}{4}$ in. Find its breadth.

2. Sizes of wires for stays are usually expressed in terms of their circumferences. Thus a $3\frac{1}{2}$ in. wire is a little over 1 in. in diameter.

$$\text{Circumference of a circle} = \tfrac{22}{7} \times \text{Diameter}$$

i.e.

$$\text{Diameter} = \text{Circumference} \div \tfrac{22}{7}.$$

EXAMPLE. What is the diameter of a $2\frac{1}{2}$ in. wire?

$$\text{Diameter} = \text{Circumference} \div \tfrac{22}{7}$$
$$= 2\tfrac{1}{2} \div \tfrac{22}{7} = \tfrac{5}{2} \times \tfrac{7}{22} = \tfrac{35}{44} \text{ in.}$$

What are the diameters of the following wire stays?

(*a*) $3\frac{1}{2}$ in. (*b*) $1\frac{1}{4}$ in. (*c*) $2\frac{1}{4}$ in. (*d*) 1 in.

3. A watch was 20 sec. slow at noon on 10 March and $8\frac{4}{5}$ sec. slow at noon on 18 March. Find the daily rate of the watch and express it as gaining or losing.

4. A watch was $13\frac{4}{5}$ sec. fast at 8.30 p.m. on 5 April and $20\frac{1}{5}$ sec. fast at 9.30 p.m. on 8 April. Find the daily rate of the watch to the nearest $\frac{1}{5}$ sec.

5. If a watch has a daily rate of $2\frac{4}{5}$ sec. gain and at noon on 1 June it is $29\frac{3}{5}$ sec. slow, when will it show correct time?

6. There is a type of fractional problem in which the answer appears to be wrong, unless the process is clearly understood.

EXAMPLE. How many pieces of string $3\frac{1}{2}$ in. long can be cut from a length of $7\frac{1}{4}$ in. and how much remains?

The answer to this is obviously 2, with $\frac{1}{4}$ in. left over.

By working with division of fractions we have

$$\frac{7\frac{1}{4}}{3\frac{1}{2}} = \frac{29}{4} \div \frac{7}{2} = \frac{29}{\underset{2}{\cancel{4}}} \times \frac{\cancel{2}}{7} = \frac{29}{14} = 2\tfrac{1}{14}.$$

The $\frac{1}{14}$ left over does *not* mean that $\frac{1}{14}$th of an inch is left, but $\frac{1}{14}$th of the *length* that is being cut off (the divisor), i.e.

$$\frac{1}{14} \text{ of } 3\tfrac{1}{2} = \frac{1}{\underset{2}{\cancel{14}}} \times \frac{\cancel{7}}{2} = \frac{1}{4} \text{ in.}$$

How many strips of aluminium each $4\frac{5}{8}$ in. long can be cut from a strip $32\frac{1}{2}$ in. long and what length remains?

7. How many pieces of Balsa wood $3\frac{1}{2}$ in. long can be cut from a length 1 ft. 8 in. and how much will be left over? Allow in each cut $\frac{1}{16}$ in. for wastage due to saw-cut.

8. Given the capacity of a petrol tank in gallons, and the hourly consumption of petrol at a certain speed, to find the possible flying time while retaining a reserve of petrol.

Example. Tank capacity 350 gallons. Petrol reserve $\frac{1}{7}$ of capacity. Consumption 40 gal. per hour.

Usable petrol $= 350 - \frac{1}{7}$ of $350 = 350 - 50 = 300$ gal.
Flying time $= 300 \div 40 = 7\frac{1}{2}$ hours.

Find the flying time in the following cases:

	Tank capacity	Reserve	Hourly consumption
(*a*)	240 gal.	$\frac{1}{8}$	30 gal.
(*b*)	216 gal.	$\frac{1}{12}$	$24\frac{1}{2}$ gal.
(*c*)	325 gal.	$\frac{1}{10}$	$20\frac{1}{4}$ gal.

9. What is the hourly consumption of petrol in the following cases if the reserve is reached in each case?

	Tank capacity	Reserve	Flying time
(*a*)	265 gal.	$\frac{1}{5}$	$5\frac{1}{2}$ hours
(*b*)	312 gal.	$\frac{1}{12}$	$7\frac{1}{4}$ hours
(*c*)	375 gal.	$\frac{1}{10}$	$8\frac{1}{3}$ hours

10. If a pump can deliver $25\frac{1}{2}$ gal. of petrol in $2\frac{1}{4}$ min., how long will it take to refuel a plane with 300 gal.?

Fractions of Concrete Quantities.

Example. What fraction is $2\frac{1}{2}$ minutes of 1 hour?

Rule. Place the quantity immediately following the word "of" in the denominator, thus: $\frac{\qquad}{1 \text{ hour}}$.

Place the other quantity in the numerator, thus: $\frac{2\frac{1}{2} \text{ minutes}}{1 \text{ hour}}$.

Bring both to the same units, i.e. minutes:

$$\frac{2\frac{1}{2} \text{ minutes}}{1 \text{ hour}} = \frac{5}{2} \times \frac{1}{60} = \frac{1}{24}.$$

Exercise X.

1. (*a*) What fraction is $12\frac{1}{2}$ min. of 1 hour?
(*b*) What fraction is $2\frac{1}{4}$ gal. of 30 gal.?
(*c*) What fraction of 110 miles is $80\frac{1}{2}$ miles?
(*d*) What fraction is 1 min. 35 sec. of 4 min. 20 sec.?
(*e*) What fraction is 2 min. 5 sec. of $8\frac{1}{2}$ min.?
(*f*) What fraction of 1 mile is 1 inch?

2. The Representative Fraction of a map (called R.F.) is an example of "fraction of a concrete quantity". It is one way of stating the scale of the map.

Suppose two aerodromes ⊙ *A* ... ⊙ *B* are 1 inch apart on a map, while on the ground they are actually 2 miles apart. What fraction is the map distance of the ground distance?

$$\text{R.F.} = \frac{\text{Map distance}}{\text{Ground distance}} = \frac{1 \text{ in.}}{2 \text{ miles}} = \frac{1}{2 \times 1760 \times 36} = \frac{1}{126{,}720}.$$

This is the R.F. of that particular map and means:

Map distance 1 in.	represents	Ground distance		126,720 in.
,, ,, 1 ft.	,,	,,	,,	126,720 ft.
, ,, 1 cm.	,,	,,	,,	126,720 cm.

The distance on a map between a light-vessel moored off a coast and charted $\underset{\smile}{*}$ from an aerodrome $\odot$ is 3 in. The actual distance apart is 15 miles. What is the R.F. of the map?

3. Find the R.F. in these cases:

	Map distance	Ground distance
(*a*)	3½ in.	3½ miles
(*b*)	8 in.	1⅓ miles
(*c*)	⅛ in.	4 furlongs

4. Express these R.F.s as miles per inch:

(*a*) $\frac{1}{21{,}120}$. (*b*) $\frac{1}{158{,}400}$. (*c*) $\frac{1}{88{,}704}$.

5. A plane carries 160 gal. of petrol and uses 35 gal. per hour at cruising speed. What fraction of the original supply is left after 2½ hours flight at this speed?

6. A plane has tank capacity 180 gal. and uses 30 gal. per hour flying at 210 m.p.h. It also uses 35 gal. per hour flying at 228 m.p.h. What fraction of the tank capacity remains after 20 min. flying time at each speed?

7. A plane's tank capacity is 120 gal., of which ⅛ is reserve. When cruising she uses 35 gal. per hour.

(*a*) What fraction is the hourly consumption of the unreserved fuel?

(*b*) How many hours cruising will this fuel load allow?

8. What fraction of a nautical mile (6080 ft.) is a statute mile (1760 yd.)?

9. On a ruler 10 in. are equal to 25⅖ cm. What fraction is 1 cm. of 1 in.?

10. A plane travelling at 220 m.p.h. increases speed by ¼. What is the new speed?

11. By what fraction of this new speed must the plane in Question 10 reduce speed to return to the original 220 m.p.h.?

12. A plane starting with a fuel load of 360 gal. uses ⅝ of it. What weight of petrol remains if 1 gal. weighs 7⅕ lb.?

DECIMALS

Addition and Subtraction.

In adding or subtracting decimals the points must be kept always under one another.

Thus 12·88 + 1·096 + 10·9 + 0·87.

```
 12·88
  1·096
 10·9
  0·87
 ------
 25·746
```

And 241·091 − 12·87.

```
 241·091
  12·87
 -------
 228·221
```

EXERCISE XI.

1. Add:

(*a*) 0·568 + 19·38 + 1·163 + 0·2218.
(*b*) 30·157 + 5·04 + 0·786 + 19·2.
(*c*) 2·0508 + 0·77 + 70·908 + 11·8903.
(*d*) 87·5 + 8·75 + 0·875 + 0·0875 + 0·00875.
(*e*) 1·98 + 18·694 + 246·224 + 0·4 + 17·935.
(*f*) 0·05 + 0·734 + 0·9049 + 0·0007 + 0·0616.
(*g*) 126·67 + 3·4716 + 10 + 81·2001 + 69·749 + 8·8 + 80·08.
(*h*) 192·8 + 17·96 + 2·3 + 5·841 + 121·3 + 64·87.
(*i*) 0·415 + 2·896 + 39·771 + 15·875 + 21·81.
(*j*) 12·987 + 1·789 + 189·92 + 16·713 + 54·787 + 0·01735.

2. Subtract:

(*a*) 21·08 − 14·89. (*b*) 13·34 − 8·891. (*c*) 12·214 − 7·006.
(*d*) 9 − 3·876. (*e*) 5·302 − 3·479. (*f*) 4·09 − 3·8654.
(*g*) 21·08 − 14·890. (*h*) 98·65 − 9·752. (*i*) 40·47 − 4·047.
(*j*) 0·212121 − 0·12121. (*k*) 1·00001 − 0·999.

3. The pressure of the atmosphere is measured by a barometer in inches of mercury. The normal pressure for the British Isles is 29·92 in. of mercury. If on a certain day the barometer indicates 31·11 in. (which is the highest recorded reading for these islands), how much is the highest above normal?

4. The lowest recorded barometer reading is 27·33. How many inches is this below normal?

5. In meteorology, or weather forecasting, the atmospheric pressure is often expressed in millibars instead of inches of mercury. The normal millibar reading is 1013·2 mb. for the British Isles. State the pressure in millibars when it is (*a*) 32·6 mb. above normal, (*b*) 82·2 mb. below normal.

6. Rainfall is measured in inches. These are the monthly totals for a certain year at Kirkwall:

4·19; 3·55; 3·45; 1·96; 1·76; 1·93; 2·90; 3·16; 3·23; 4·74; 4·57; 4·64.

Find (*a*) total annual rainfall, (*b*) difference between greatest and least month's rain.

Decimal Multiplication.

EXAMPLE. Multiply 12·42 by 3·6.

First multiply without points:

$$\begin{array}{r} 1242 \\ 36 \\ \hline 7452 \\ 3726 \\ \hline 44712 \end{array}$$

Count the total number of figures *after* the decimal points in the two quantities to be multiplied. In this case, 3.

Count 3 figures from the right in the answer and insert the decimal point, i.e.

Answer is 44·712.

EXERCISE XII.

1. By inspection state the number of figures after the decimal point in these answers:

(*a*) 3·14 × 12·206. (*b*) 2·001 × 1·8. (*c*) 0·00168 × 0·07.
(*d*) 144 × 0·02. (*e*) 5·06 × 0·0001. (*f*) 0·01 × 0·00205.

2. Find the product of:

(*a*) 2·55 × 2. (*b*) 80·6 × 5. (*c*) 9·8 × 9. (*d*) 5·3 × 8.
(*e*) 0·682 × 11. (*f*) 0·0162 × 9. (*g*) 2·93 × 12. (*h*) 0·0492 × 11.

3. Simplify:

(*a*) 4·24 × 1·4. (*b*) 21·4 × 0·013. (*c*) 216·2 × 2·2.
(*d*) 0·013 × 0·9. (*e*) 4·6 × 43·02. (*f*) 9·8 × 0·0803.
(*g*) 124·8 × 0·08. (*h*) 124·876 × 21·5. (*i*) 3·142 × 9·78.
(*j*) 730 × 0·00609. (*k*) 4·296 × 0·0843. (*l*) 0·7261 × 75·4.
(*m*) 0·0033 × 6·507. (*n*) 9·191 × 1·919.

4. Find the area of these rectangular plates (Area = Length × Breadth):

(*a*)		(*b*)		(*c*)	
Length	Breadth	Length	Breadth	Length	Breadth
3·25 in.	2·12 in.	4·25 in.	3·25 in.	7·63 in.	3·07 in.

5. 1 millibar = 0·029 in. of mercury. Convert the following millibar readings to barometric inches:

(*a*) 1007·2 mb. (*b*) 996·4 mb. (*c*) 1041 mb.

Answer to first decimal place.

6. The International Commission for Air Navigation (I.C.A.N.) uses 15° C. as normal ground temperature. This decreases by 1·98° C. for every 1000 ft. rise in altitude up to 36,090 ft. Above this it remains unchanged. If the ground temperature is normal, find the temperatures at these altitudes:

(*a*) 4000 ft. (*b*) 5000 ft. (*c*) 10,000 ft. (*d*) 36,000 ft.

7. Atmospheric pressure in low altitudes falls 1 millibar for every 30 ft. increase in height. If the barometric reading is I.C.A.N. normal at ground level, i.e. 1013·2 mb., what are the readings at these altitudes?

(*a*) 6000 ft. (*b*) 7200 ft. (*c*) 12,000 ft.
(*d*) 5400 ft. (*e*) 3300 ft. (*f*) 1950 ft.

8. Find the altitudes at which the following readings were obtained:

(*a*) 780·2 mb. (*b*) 650 mb. (*c*) 926·4 mb.
(*d*) 900 mb. (*e*) 862·2 mb.

Decimal Division.

First convert the divisor into a whole number. Thus, in

$$16{\cdot}842 \div 4{\cdot}2,$$

move the decimal point in 4·2 one place to the right, i.e. 42.

Equalise this by moving the decimal point in the dividend one place to the right, i.e. 168·42.

Place the decimal point in the answer immediately after bringing down the first figure of the decimal, i.e. 4. Thus

```
     4·01
   ------
42)168·42
   168
   ---
      ·42
       42
       --
       ..
```

EXAMPLE. 151·262 ÷ 3·61 becomes 15126·2 ÷ 361.

```
        41·9
    --------
361)15126·2
    1444
    ----
      686
      361
      ---
     3252
     3249
     ----
        3
```

The division is not complete, as there is still a remainder. But 15126·2 = 15126·2000·····, so as many noughts as required can be brought down.

If the answer is to be to three decimal places, the division is continued:

```
        41·900
361)15126·2
    1444
     686
     361
     3252
     3249
        300
```

The answer to three decimal places is 41·900.

If the *nearest* third decimal place is wanted the division must be carried to four places. If this fourth figure is 5 or over we add 1 to the third place.

In the above example the fourth figure of decimals is 9. Therefore the answer to the nearest *third* decimal place is 41·901.

EXERCISE XIII.

1. Simplify:

(*a*) 52·5 ÷ 2·5. (*b*) 28·08 ÷ 0·9. (*c*) 1·44 ÷ 1·2.
(*d*) 15·21 ÷ 1·17. (*e*) 15·786 ÷ 0·06. (*f*) 0·0062 ÷ 2·5.
(*g*) 183·82 ÷ 0·091. (*h*) 13·014 ÷ 2·41. (*i*) 264·708 ÷ 3·24.
(*j*) 28·782 ÷ 3·69. (*k*) 906·5 ÷ 0·185. (*l*) 1769·08 ÷ 0·47.
(*m*) 91·008 ÷ 379·2. (*n*) 0·4496 ÷ 11·24.
(*o*) 0·119385 ÷ 22·74. (*p*) 542·913 ÷ 0·05271.

2. Express as decimal to number of places stated:

(*a*) 0·51 ÷ 6·25 (to 3rd pl.). (*b*) 2·653 ÷ 7·4 (to 3rd pl.).
(*c*) 87 ÷ 426·8 (to 2nd pl.).
(*d*) 12·85 ÷ 17·2 (to nearest 3rd pl.).
(*e*) 0·014 ÷ 2·3 (to 2nd pl.). (*f*) 0·00565 ÷ 9·4 (to 5th pl.).
(*g*) 5·42913 ÷ 527·1 (to nearest 3rd pl.).
(*h*) 1 ÷ 846 (to nearest 4th pl.).
(*i*) 186·74 ÷ 53·2 (to nearest whole number or "integer" as it is called).
(*j*) 10·7358 ÷ 0·174 (to nearest integer).

3. The area of a rectangular plate is 17·8932 sq. in. The breadth is 3·72 in. What is the length?

4. Any full circle contains 360 degrees (360°) and each degree contains 60 minutes (60′). Thus a full circle contains 21,600 minutes.

The length of 1 minute of the earth's equator is called a Nautical Mile, i.e. the equator is 21,600 nautical miles in length. What is the diameter of the earth in nautical miles? (Circumference = 3·14 × diameter.)

5. On a model globe 10° of the equator measure 1·22 in. Find the diameter of this globe to nearest first decimal place.

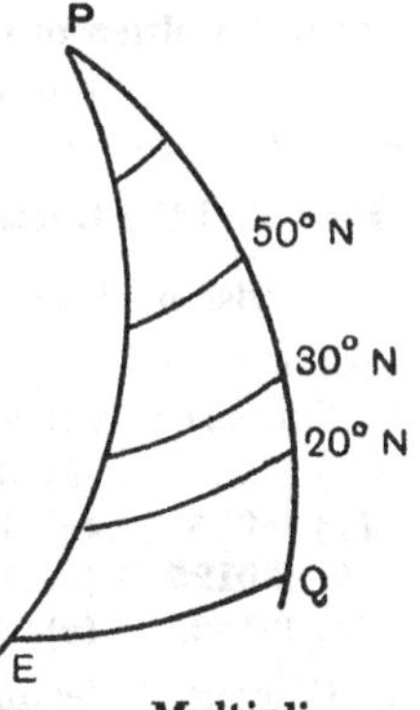

6. *P* represents the true N. pole of the earth and *EQ* a portion of the equator. It is clear from the diagram that the two meridians *PE* and *PQ* get closer together the nearer they get to the pole, until they meet at the pole. The distance they are apart at any latitude can be found by multiplying *EQ* by a decimal factor

The table below gives the actual nautical miles, at different latitudes, between two meridians 60 nautical miles (1°) apart at the equator. Find the multiplying factor.

	Lat. N.	Nautical miles	Multiplier
Equator	0°	60	
(*a*)	10°	59·09	
(*b*)	30°	52·00	
(*c*)	50°	38·64	
(*d*)	70°	20·58	

7. Find the total resistance of two wires of 2·4 ohms and 3·6 ohms resistance respectively when joined in parallel. (See Question 4, Ex. V.)

8. What is the total resistance of three wires in parallel whose resistances are 1·8 ohms, 1·2 ohms and 2·7 ohms?

9. How many lengths of wire each measuring 19·7 ft. can be cut from a reel containing 1000 yd. and what length remains? (See Question 6, Ex. IX.)

10. How many lengths of Balsa wood 7·5 in. long can be cut from a length 10 ft. 9 in. and what length remains? Allow 0·07 in. wastage for each piece due to saw cut.

Conversion of Decimals to Vulgar Fractions.

Example. Convert 0·013 to a vulgar fraction.

Place all the useful figures after the decimal point as the numerator of the fraction.

For the denominator place 1 for the decimal point and as many noughts as there are figures after the decimal point (*including* noughts this time). Thus

$$0{\cdot}013 = \tfrac{13}{1000}.$$

Conversion of Vulgar Fractions to Decimals.

Example. Change $\frac{3}{4}$ to a decimal.

$\frac{3}{4}$ means 3 divided by 4. Do it.

$$\tfrac{3}{4} = 0{\cdot}75.$$

Decimal Values of Concrete Quantities.

EXAMPLE. What decimal is a cruising speed of 155 m.p.h. of a maximum speed of 185 m.p.h.?

First find the fraction $\frac{\text{Cruising speed}}{\text{Maximum speed}}=\frac{155}{185}=\frac{31}{37}$.

By division, this $=0{\cdot}837$.

EXERCISE XIV.

1. Convert to vulgar fractions in their lowest terms:

(*a*) 0·5	(*b*) 0·95.	(*c*) 0·15.	(*d*) 0·94.	(*e*) 0·875.
(*f*) 0·075	(*g*) 1·28.	(*h*) 5·65.	(*i*) 2·45.	(*j*) 1·025.
(*k*) 2·0125.	(*l*) 10·001.	(*m*) 3·504.	(*n*) 10·375.	(*o*) 0·0256.
(*p*) 3·142.	(*q*) 6·0625.	(*r*) 2·044.	(*s*) 1·408.	(*t*) 0·0011.

2. Convert to decimals (to three places):

(*a*) $\frac{6}{7}$. (*b*) $\frac{5}{8}$. (*c*) $\frac{5}{16}$. (*d*) $\frac{6}{13}$. (*e*) $\frac{17}{18}$. (*f*) $\frac{23}{32}$.

3. What decimal of a full load of 300 gal. of petrol is used in 2 hours' flight consuming 55 gal. per hour?

4. If the maximum speeds of a Miles Magister and D.H. Moth Minor are 145 m.p.h. and 118 m.p.h. respectively, what decimal is the second speed of the first?

5. What decimal is 1 yard of 1 metre, if 1 metre $=39{\cdot}37$ in.?

6. What decimal of 1 nautical mile (6080 ft.) is one statute mile (1760 yd.)? Answer to third decimal place.

7. What decimal is 1° F. of 1° C.?

8. If 29·92 in. mercury of barometric pressure $=1013{\cdot}2$ millibars and 31·11 in. mercury of barometric pressure $=1053{\cdot}5$ millibars, what decimal is a millibar of a "mercury" inch?

COMPLEX FRACTIONS AND DECIMALS

These must be worked in accordance with the "Order of Signs" as shown in the following table:

(*a*) Brackets must be cleared first.

(*b*) Quantities connected by "of" next.

(*c*) $\times$ and $\div$ are equal in value and are taken in their order of occurrence.

(*d*) $+$ and $-$ are also equal and come last, in order of occurrence.

NOTE. Squares and square root signs both act as brackets.

When a fraction is complex in both the numerator N and denominator D, they should be worked separately.

EXAMPLE. Simplify: $\dfrac{\frac{3}{4}+1\frac{2}{3} \text{ of } 1\frac{1}{5}-\frac{2}{5}\div 2\frac{2}{3}\times 1\frac{1}{8}}{(\frac{3}{4}+1\frac{2}{3})\times 1\frac{1}{5}-\frac{2}{5}\div 2\frac{2}{3} \text{ of } 1\frac{1}{8}}$.

N

$$\frac{3}{4}+\frac{5}{3} \text{ of } \frac{6}{5}-\frac{2}{5}\div 2\tfrac{2}{3}\times 1\tfrac{1}{8}$$

$$=\frac{3}{4}+2-\frac{2}{5}\times\frac{3}{8}\times\frac{9}{8}$$

$$=\frac{3}{4}+2-\frac{27}{160}$$

$$=2\frac{120-27}{160}=2\tfrac{93}{160}.$$

D

$$\left(\frac{3}{4}+\frac{5}{3}\right)\times 1\tfrac{1}{5}-\frac{2}{5}\div 2\tfrac{2}{3} \text{ of } 1\tfrac{1}{8}$$

$$=\left(\frac{9+20}{12}\right)\times\frac{6}{5}-\frac{2}{5}\div\frac{8}{3} \text{ of } \frac{9}{8}$$

$$=\frac{29}{12}\times\frac{6}{5}-\frac{2}{5}\times\frac{1}{3}$$

$$=\frac{174}{60}-\frac{2}{15}=\frac{174-8}{60}=\frac{166}{60}.$$

The fraction then becomes

$$\frac{N}{D}=\frac{2\frac{93}{160}}{\frac{166}{60}}=\frac{413}{160}\times\frac{60}{166}=\frac{1239}{1328}.$$

EXERCISE XV.

Simplify the following:

1. $\frac{1}{2}+\frac{2}{3}$ of $\frac{2}{5}+\frac{3}{4}$.
2. $(\frac{1}{2}+\frac{2}{3})$ of $\frac{2}{5}+\frac{3}{4}$.
3. $(\frac{1}{2}+\frac{2}{3})$ of $(\frac{2}{5}+\frac{3}{4})$.
4. $\frac{1}{2}+\frac{2}{3}$ of $(\frac{2}{5}+\frac{3}{4})$.
5. $\frac{3}{4}\div 1\frac{1}{2}\times 1\frac{1}{3}$.
6. $\frac{3}{4}\div 1\frac{1}{2}$ of $1\frac{1}{3}$.
7. $1\frac{1}{2}+1\frac{2}{3}-\frac{1}{5}$ of $1\frac{7}{8}-(\frac{3}{4}-\frac{5}{8})$.
8. $\dfrac{2\frac{1}{2}+\frac{1}{3} \text{ of } 1\frac{1}{5}-\frac{3}{4}}{2\frac{1}{2}+\frac{1}{3} \text{ of } (1\frac{1}{5}-\frac{3}{4})}$.
9. $\dfrac{(1\frac{2}{3}-\frac{3}{4})\div 1\frac{1}{2}\times\frac{3}{8}}{1\frac{2}{3}-\frac{3}{4}\div 1\frac{1}{2} \text{ of } \frac{3}{8}}$.
10. $\dfrac{2\frac{1}{3}-\frac{3}{4}\div\frac{3}{5}+\frac{2}{3}\times\frac{7}{8}}{(2\frac{1}{3}-\frac{3}{4})\div(\frac{3}{5}+\frac{2}{3})\times\frac{7}{8}}$.

EXERCISE XVI.

Simplify, giving answers to three decimal places:

1. $4\cdot 5\div 2\cdot 25\times 10\cdot 8$.
2. $4\cdot 5\div 2\cdot 25$ of $10\cdot 8$.
3. $(3\cdot 1+2\cdot 2+0\cdot 16+1\cdot 44)\times(1\cdot 4-1\cdot 38)$.
4. $(3\cdot 6-1\cdot 02\times 2\cdot 8)\div 0\cdot 6$ of $1\cdot 2$.
5. $12\cdot 8\div 3\cdot 2\times 1\cdot 5-1\cdot 75$.
6. $12\cdot 8\div 3\cdot 2$ of $1\cdot 5-1\cdot 75$.
7. $9\cdot 6-2\cdot 4$ of $(2\cdot 8-1\cdot 35)$.
8. $\dfrac{2\cdot 1-1\cdot 5\div(3\times 0\cdot 25)}{2\cdot 1+1\cdot 6\div 0\cdot 8-0\cdot 1}$.
9. $\dfrac{2\cdot 1-\dot{1}\cdot 2\div 0\cdot 3\times 0\cdot 25}{2\cdot 1-1\cdot 5\div 0\cdot 8 \text{ of } 2\cdot 5}$.
10. $\dfrac{(1\cdot 1+2\cdot 2)-0\cdot 5(2\cdot 2-1\cdot 1)}{1\cdot 1-2\cdot 2\times 0\cdot 5\div 2\cdot 5}$.

RELATIVE SPEEDS

When one plane is overtaking another on the same course, the actual gaining speed is the difference of the two speeds. This is termed the "relative speed" of the two planes.

EXAMPLE 1. At noon, a pursuing plane *A*, flying at 105 m.p.h., is 30 miles astern of plane *B* flying at 95 m.p.h. At what time will *A* overtake *B*?

Difference in speeds (relative speed) $= 105 - 95 = 10$ m.p.h.

Therefore the gaining speed of $A = 10$ m.p.h. and time taken to cover 30 miles $= \frac{30}{10} = 3$ hours.

Therefore time that *A* overtakes *B* is 3 p.m.

If we desire to know the actual distances covered by the two planes during the interval we must multiply the time interval by the actual speeds of the planes, i.e.

A will have travelled $3 \times 105 = 315$ miles,
B „ „ $3 \times 95 = 285$ miles,

from which it will be seen that *A* has flown 30 miles further than *B*.

If two planes are approaching one another the relative speed is obtained by adding the two speeds.

EXAMPLE 2. Two planes *A* and *B* are 120 miles apart approaching each other. *A* is flying at 100 m.p.h. and *B* at 80 m.p.h. How long will they take to meet?

Relative speed $= 100 + 80 = 180$ m.p.h.

Time taken to fly 120 miles at 180 m.p.h.

$= \frac{120}{180} = \frac{2}{3}$ hour or 40 min.

During this time *A* will have flown $\frac{2}{3} \times 100 = 66\frac{2}{3}$ miles
and *B* will have flown $\frac{2}{3} \times 80 = 53\frac{1}{3}$ miles
which together total 120 miles (the distance they were apart).

EXERCISE XVII.

1. A plane is flying "down-wind" (i.e. with wind astern) and its ground speed is 240 m.p.h. One quarter of this is due to wind speed. What fraction will the "up-wind" (or return) ground speed be of the outward ground speed?

2. At 10 a.m. plane *A*, flying at 180 m.p.h., is 4 miles behind, and overtaking plane *B* flying at 150 m.p.h. At what time will *A* overtake *B*?

3. Two planes are patrolling between stations *A* and *B*, 50 miles apart. The plane leaving *A* flies at 80 m.p.h. and the one leaving *B* flies at 100 m.p.h. Both leave their respective stations at the same time. After how long will they meet?

4. Plane A, flying at 120 m.p.h., is overtaking plane B flying at 100 m.p.h. At noon A is 5 miles astern of B. Find (*a*) at what time A will open fire if her opening range is 440 yd., (*b*) how far plane A will have flown between noon and that time.

5. A plane flying down-wind with air speed indicator showing 90 m.p.h. is on a course to meet another plane flying up-wind with air speed indicator also showing 90 m.p.h. If the wind speed throughout is 20 m.p.h., find (*a*) at what time they will meet if at 3.30 p.m. they are 54 miles apart, (*b*) how far each will have flown from 3.30 p.m. until they meet.

6. Two planes approach one another from stations 40 miles apart. They meet in 10 min. Find the speeds of each in m.p.h. if the first plane travelled $\frac{5}{8}$ of the total distance.

7. A plane flying at 96 m.p.h. gains 3 miles on another plane, flying in the same direction, in 12 min. What is the speed of the second plane?

8. NOTE. When the distance between two vessels is diminishing, the gun range is said to be "closing". The "Range Rate" is the number of yards per minute by which the range is changing.

The range rate is closing if distance apart is diminishing, and opening if distance apart is increasing.

For range rate purposes 1 nautical mile is usually taken as 2000 yd.

A cruiser steaming at 25 knots is overtaking a transport steaming at 19 knots. The cruiser is 10 nautical miles astern at 2 p.m. Find (*a*) at what time the gun range will be 10,000 yd., (*b*) the range rate in yd. per min.

9. A battleship steaming 20 knots is pursuing a cruiser. The range at 10 a.m. is 15,000 yd., and the range rate is "250 yd. per min. opening". Find (*a*) the speed of the cruiser, (*b*) when the cruiser will be at "extreme range", if the maximum range of the battleship's guns is 20,000 yd.

10. An aircraft carrier in position A at 11 a.m. is steaming true North at 20 knots. A seaplane leaves the carrier at 11 a.m. to scout 50 nautical miles true North and return. The ground speed of the seaplane is 100 knots. Find (*a*) the time at which the seaplane alters course to true South, (*b*) the time of return to the aircraft carrier, (*c*) the total distance travelled by the seaplane.

11. Three aeroplanes A, B and C fly a 20 mile race. A beats B by $\frac{1}{2}$ mile and C by $\frac{3}{4}$ mile. By how much would B beat C in a 78 mile race?

12. A night bomber flying at 19,620 ft. at a speed of 200 m.p.h. appears by the sound of its engines to be vertically above your head. How far from such a position has it actually flown? (Sound travels at 1090 ft. per sec.)

DECIMAL OR METRIC SYSTEM OF WEIGHTS AND MEASURES

used in most countries except the British Empire and U.S.A.

Prefixes.	milli = thousandth ($\frac{1}{1000}$ of).	
	centi = hundredth ($\frac{1}{100}$ of).	
	deci = tenth ($\frac{1}{10}$ of).	
	deca = ten times (10 ×).	
	hecto = hundred times (100 ×).	
	kilo = thousand times (1000 ×).	
Length.	Unit: 1 metre	= 39·37 inches.
	10 millimetres (mm.)	= 1 centimetre.
	10 centimetres (cm.)	= 1 decimetre.
	10 decimetres (dm.)	= 1 metre.
	10 metres (m.)	= 1 Dekametre.
	10 Dekametres (Dm.)	= 1 Hectometre.
	10 Hectometres (Hm.)	= 1 Kilometre (Km.).
Area.	100 sq. cm.	= 1 sq. decimetre.
	100 sq. dm.	= 1 sq. metre.
Volume.	Unit: 1 litre	= 1·76 pints.
	1000 cubic mm. (c.mm.)	= 1 cubic centimetre (c.c.).
	1000 c.c.	= 1 cubic decimetre (litre).
	1000 litres	= 1 cubic metre.
Weight.	Unit: 1 gram (gm.).	
	1000 grams	= 1 kilogram (kgm.) = 2·204 lb.
	1000 kilograms	= 1 tonne.

EXERCISE XVIII.

1. (i) What fraction, (ii) What decimal, is:

(*a*) 1 mm. of 1 m. (*b*) 1 m. of 1 Km. (*c*) 1 cm. of 1 m.
(*d*) 3 cm. of 1 dm. (*e*) 4 m. of 1 Km. (*f*) 40 cm. of 1 Km.?

2. Express as kilometres and decimals of a kilometre:

(*a*) 5 Km. 4 m. (*b*) 1 Km. 156 m. (*c*) 56 m.
(*d*) 750 m. (*e*) 2 m. 4 dm. (*f*) 4 m. 46 cm.

3. Convert:

(*a*) 2·462 Km. to cm. (*b*) 2613 c.c. to litres.
(*c*) 4280 gm. to kgm. (*d*) 22 litres to c.c.
(*e*) 76 cm. to metres. (*f*) 466 sq. cm. to sq. metres.
(*g*) 3 tonnes 2 kgm. to gm. (*h*) 66,750 gm. to tonnes.
(*i*) 7 c.mm. to litres. (*j*) 2·4 litres to c.c.

4. Express the following as required:
   (*a*) 7 m. 3 dm. 4 cm. + 9 dm. 3 cm. (answer in m.).
   (*b*) 42 Dm. + 724 mm. (answer in m.).
   (*c*) 125 m. 624 mm. + 71 m. 937 mm. (answer in mm.).
   (*d*) 9·24 Hm. + 16·2 m. + 42,000 mm. (answer in Km.).
   (*e*) 4 kgm. + 321 gm. − 1·6 kgm. (answer in gm.).
   (*f*) 4 cub. metres + 16·5 litres (answer in c.c.).

Conversion of Metric Measures into English Measure.

EXAMPLE. Given that 1 mile $= \frac{66}{41}$ Km., convert 182 miles into Km. (answer to one decimal place).

If 1 mile $= \frac{66}{41}$ Km., then

$$182 \text{ miles} = \frac{66 \times 182}{41} \text{ Km.} = \frac{12{,}012}{41} \text{ Km.} = 292{\cdot}9 \text{ Km.}$$

EXERCISE XIX.

1. Convert the following miles into kilometres:
   (*a*) 63. (*b*) 214. (*c*) $81\frac{1}{4}$. (*d*) 100.

2. Convert the following kilometres into miles:
   (*a*) 100. (*b*) 571. (*c*) 32·5. (*d*) 179.

3. Express a "ceiling" of 30,000 ft. in metres. (1 m. = 3·28 ft.)

4. Which has the higher service ceiling to the nearest foot?

(*a*)	V. A. Wellington Mk Ia	26,300 ft.
or	Heinkel III K Mk Va	7,392 metres.
(*b*)	A. W. Whitley IV	25,000 ft.
or	Italian Breda 88	8,543 metres.

5. Which bobbin contains the most wire and by how many feet?
   Bobbin *A* containing 67 yd.
   Bobbin *B* containing 61 metres.

6. The length of the earth's equator is 40,000 kilometres or 21,600 nautical miles. Express 1 nautical mile in kilometres.

7. What decimal (three places) is 1 Km. of 1 N.M.?

8. In Sept. 1932 Capt. Uwins in a Vickers Vespa obtained the world's altitude record (up till then) of 13,404 metres. Express this in feet.

9. What decimal is 1 cm. of 1 in.? (1 m. = 39·4 in.)

10. If 1 metre = 3·28 ft., what decimal is 1 Km. of 1 mile?

11. Express 1 sq. metre in sq. ft. (to two decimal places).

12. What decimal is 1 sq. ft. of 1 sq. metre?

Conversion of Metric and English Capacities.

EXERCISE XX.

1. Convert these litres to gallons (1 litre $=0{\cdot}22$ gal.):
(*a*) 1000. (*b*) 600. (*c*) 450. (*d*) $2\frac{1}{2}$ kilolitres.

2. Convert these gallons to litres (1 gal. $=4{\cdot}546$ litres):
(*a*) 1000. (*b*) 600. (*c*) 450. (*d*) $4\frac{1}{3}$.

3. A kilolitre tank is full but springs a leak. The loss is 100 c.c. per sec. After how many minutes will the tank be empty?

4. A plane *A* uses 42 gal. fuel per hour and plane *B* uses 178 litres per hour. Which has the greater fuel consumption and by how many gallons?

5. If 350 litres of petrol are pumped into a tank in mistake for 350 gal., how many gallons short is the measure?

6. If 350 litres of petrol are pumped into a tank of capacity 350 gal., what fraction of the tank is still available for fuel?

7. Which plane retains the greater reserve of petrol and by how many gallons?
Plane *A*. Tank capacity 450 gal. Reserve $\frac{1}{9}$th.
Plane *B*. Tank capacity 2000 litres. Reserve $\frac{1}{8}$th.

8. Which is the more efficient pump and by how many gal. per min.?
Pump *A* delivers 360 gal. in 5 min.
Pump *B* delivers 978 litres in 3 min.

9. What decimal is 250 c.c. of 1 gal.?

10. If 1 kilolitre $=35{\cdot}32$ cub. ft. and 1 gallon $=277$ cub. in., how many gallons are there in 1 kilolitre (to nearest gallon)?

Conversion of Metric and English Weights.

EXERCISE XXI.

1. A man weighs 12 st. 6 lb. What is his weight in kgm.?

2. If petrol weighs $7{\cdot}2$ lb. per gal., find the weight of 1 kilolitre in kgm.

3. Find which plane carries the heavier petrol load and by how many pounds.
Plane $A=300$ gal. Plane $B=1400$ litres.
(1 c.c. petrol weighs $0{\cdot}72$ gm.)

4. In 1903 Wright's aeroplane had a wing surface of 600 sq. ft. and weighed 925 lb. This gave a wing loading of $\frac{925}{600}=1{\cdot}54$ lb. per sq. ft. Express these facts in metric units.

5. Compare these wing loadings by conversion of metric to English units, i.e. lb. lifted per sq. ft.:

(*a*) Supermarine S 6 B racing seaplane: Wing area 145 sq. ft. Weight 5995 lb.

(*b*) Macchi 72 Racing seaplane: Wing area 15·5 sq. metres. Weight 3202·5 kgm.

6. A British standard wire rope of circumference $\frac{3}{4}$ in. has a breaking load of 1·7 tons. Express these facts in metric units, given that 1 in. = 2·54 cm. and 1 metric tonne = 2204·6 lb.

The Statute Mile and Nautical Mile.

The English Statute mile is 1760 yd. The Nautical mile is that length of the earth's equator which subtends an angle of 1 minute at the earth's centre, i.e.

The nautical mile $= \frac{1}{21,600}$ of the earth's equator.

The length of the nautical mile = 6080 ft.;

$$\therefore \quad \frac{\text{Nautical mile}}{\text{Statute mile}} = \frac{608\emptyset}{528\emptyset} = \frac{76}{66} = \frac{38}{33}.$$

Therefore 1 N.M. = $\frac{38}{33}$ of 1 S.M.

and 1 S.M. = $\frac{33}{38}$ of 1 N.M.

Also 1 nautical mile is the same as 1 minute of latitude.

Thus, if a plane travels 63 nautical miles due North from the equator it is then in latitude 63′ N., i.e. 1° 3′ N. lat.

Statute miles must be converted to nautical miles to obtain latitude.

EXERCISE XXII. (Answer in each case to nearest mile.)

1. Convert the following to statute miles:

(*a*) 214 N.M. (*b*) 650 N.M. (*c*) 15·2 N.M. (*d*) 2012 N.M. (*e*) 576·8 N.M.

2. Change to nautical miles:

(*a*) 204 S.M. (*b*) 625 S.M. (*c*) 17·4 S.M. (*d*) 1990 S.M. (*e*) 562·7 S.M. Answers to first decimal place.

Latitude and distance covered in True N. or S. direction.

The equator is lat. 0 and places not on the equator may have either North (N.) or South (S.) latitude. Care must be taken to define which it is.

An aircraft in N. lat. travelling North is increasing its latitude, while if travelling South the latitude is decreasing.

Similarly in S. lat., an aeroplane going South is increasing latitude and going North is decreasing latitude.

Crossing the equator involves changing latitude from N. to S. or vice versa.

The point from which an aircraft leaves or takes off is known as a "departure point". The place the aircraft reaches is called the "arrival point".

EXAMPLE. A plane travels 212 statute miles on a course N. true from a departure point lat. 15° 49′ N. Find the new latitude.

First correct statute miles to nautical miles:

$$1 \text{ S.M.} = \tfrac{33}{38} \text{ N.M.}$$

$$\therefore \text{ Distance travelled} = 212 \text{ S.M.} = \frac{212 \times 33}{38} \text{ N.M.} = 184{\cdot}1 \text{ N.M.}$$

Thus difference of lat. = 184·1 min. of N. lat. = 3° 4·1′ N. lat.

∴ New lat. = (15° 49′ + 3° 4·1′) N. = 18° 53·1′ N.

EXERCISE XXIII.

Find the latitude of arrival in the following (to nearest min.):

	Course	Distance in S.M.	Lat. of Dep. Point
1.	N. true	386	20° 30′ N.
2.	S. true	274	1° 15′ N.
3.	N. true	624	3° 57′ S.
4.	N. true	418	52° 24′ N.
5.	S. true	1346	50° 31′ N.
6.	N. true	75	39° 59′ S.

Statute distances from change of latitude when travelling True N. or S.

EXAMPLE. A plane has to fly N. true from departure point 50° 20′ N. to lat. 55° 31′ N. How many statute miles must the plane travel?

Departure point 50° 20′ N.
Arrival point 55° 31′ N.

Difference of latitude = 5° 11′ N. = 311 min. or nautical miles

$$= \tfrac{38}{33} \times 311 \text{ statute miles} = 358 \text{ S.M.}$$

EXERCISE XXIV.

Find the distance in statute miles in the following (to nearest mile):

	Lat. of departure	Lat. of arrival	Course
1.	32° 38′ N.	30° 5′ N.	S. True
2.	52° 24′ N.	57° 36′ N.	N. True
3.	34° 18′ S.	31° 6′ S.	N. True
4.	2° 4′ N.	3° 18′ S.	S. True
5.	8° 15′ S.	1° 49′ N.	N. True
6.	64° 10′ N.	60° 12′ N.	S. True

AVERAGES

Suppose a reconnaissance plane, on four separate days, to fly the following distances in miles: 212·8; 327·6; 954·4; 706.

The average distance flown per day is found by totalling the four distances and dividing by the number of days, i.e.

$$\begin{array}{r} 212{\cdot}8 \\ 327{\cdot}6 \\ 954{\cdot}4 \\ \underline{706{\cdot}0} \\ 4)\underline{2200{\cdot}8} \\ 550{\cdot}2 \text{ miles per day.} \end{array}$$

Exercise XXV.

1. Find the average daily mileage from the following miles per day: (*a*) 112·3; 214·2; 326; 483·4. (*b*) 235·7; 61; 189·7; 238·6; 134.

2. For six days a plane averaged 152·2 miles per day. What was the total mileage for the six days?

3. For four days the average mileage of a plane was 210·2 miles per day. For the first three days the actual mileage covered was 260; 148; 176. What was the distance flown on the fourth day?

4. The average speed of a plane for a 3 hours flight was 210 m.p.h. Her speed during the first hour was a steady 190 m.p.h. and for the second hour 203 m.p.h. Find the speed for the third hour.

5. An aircraft flew at 180 m.p.h. for 2 hours and 196 m.p.h. for 3 hours. What was the average speed in m.p.h.?

6. In May 1928 Sir C. Kingsford Smith, in the "Southern Cross", covered 7800 miles in 81 hours 19 min. actual flying time. Find his average speed in m.p.h.

7. A pilot flew a distance of 552½ miles at a speed of 221 m.p.h. and then increased speed to 240 m.p.h. for a further 840 miles. What was his average speed (to nearest m.p.h.)?

8. In Question 7 the pilot flew the first 552½ miles at a height of 2000 ft. and then increased to 2800 ft. for the rest of the journey. What was his average altitude (to nearest 20 ft.)?

9. Find the average length of the following bundle of wires: 3 of length 4 ft. 6 in.; 5 of length 5 ft. 3 in.; 8 of length 8 ft. 6 in.

10. What is the average of the following sextant readings? 31° 15′ 10″; 31° 16′ 0″; 31° 14′ 50″; 31° 15′ 20″.

11. The following daily record of atmospheric pressure in millibars was recorded during a week. Find the average for the week expressed in inches of mercury (1 mb. = 0·029 in. mercury): 1001·6; 1006·2; 1008·7; 1017·6; 1004·8; 1002·4; 1000·1.

12. Contour lines on a map are lines joining points of equal altitude above sea level.

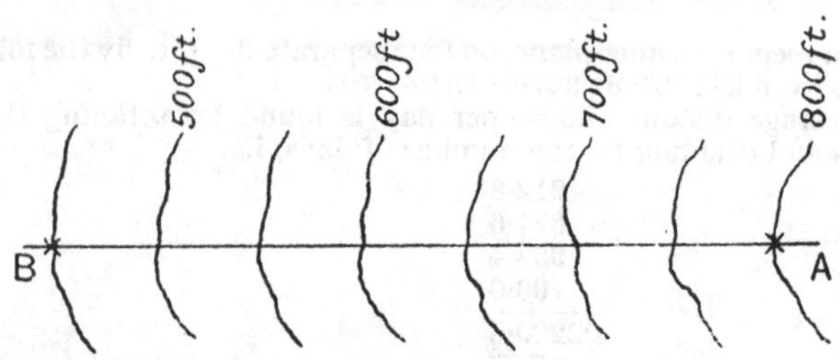

A plane is flying from A to B and the altimeter shows a steady height of 2000 ft. The contours are equally spaced. Find the average height of the plane above ground.

(NOTE. An altimeter records height above mean sea level and not above ground.)

13. Two aerodromes A and B are at equal altitudes and 80 miles apart. At A the barometer shows 1018·2 mb. and at B 1001·8 mb. Find the average pressure gradient (i.e. rate of change) in millibars per mile from A to B.

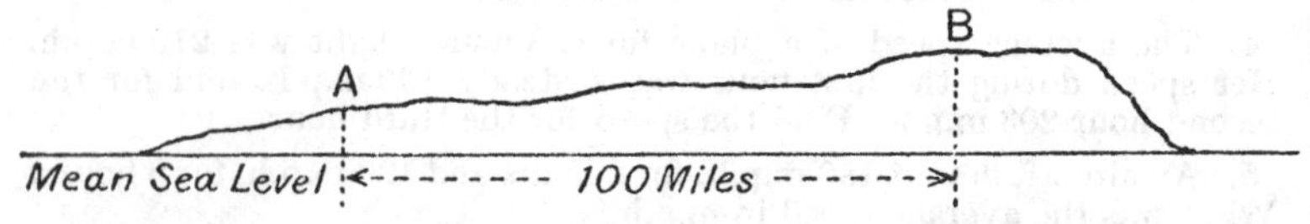

14. Aerodrome A is 150 ft. above mean sea level and the barometer reads 1010·4 mb. Aerodrome B is 360 ft. above M.S.L., and if weather conditions are the same at both places, what would be the barometer reading at B? (1 mb. = 30 ft. rise in altitude.)

15. In Question 14 assume that the weather conditions are different at B and the pressure is 1012·8 mb. By reducing both to mean sea level readings and thus eliminating the effect of difference of altitude, find the pressure gradient from A to B in millibars per mile.

HARDER PROBLEMS FOR REVISION

EXERCISE XXVI.

Assume the following data where necessary:

41 statute miles = 66 kilometres.
1 nautical mile = $\frac{38}{33}$ statute miles.
1 gallon of petrol weighs 7·2 lb.
1 millibar = 0·029 in. of mercury.
1 millibar = 30 ft. rise of altitude.

Fall in temp. per 1000 ft. rise of altitude = 1·98° C.

1. A Spitfire flying at 360 m.p.h. is 1080 yd. behind a Dornier flying at 418·5 Km. per hour. How long (to the nearest sec.) will it take the pilot of the Spitfire to close to 200 yd. to open fire?

2. The ground reading of atmospheric pressure is 1010 mb. The barometric reading in an aeroplane is 670 mb. What is the temperature in ° C. at this height if ground temperature is 20·19° C.?

3. The ground reading of atmospheric pressure is 30·16 in. of mercury. The reading in an aeroplane is 583 mb. What is the temperature in ° C. at this height if ground temperature is 17·5° C.?

4. The ground reading of atmospheric pressure is 29·58 in. of mercury. The barometric pressure in an aeroplane is 492 mb. What is the temperature in ° F. at this altitude if the ground temperature is 19·4° C.?

5. A plane flies due N. from departure point lat. 5° 41′ N. with air speed 250 statute miles per hour. A wind is blowing throughout from due N. at 20 knots. What is the latitude of arrival point after 3 hr. 15 min. flying?

6. A plane flies due N. from departure point 10° 21′ N. lat. for 2 hr. 30 min. and then turns and flies back due S. for 4 hr. 10 min. at constant air speed of 380 statute miles per hour. A wind blows throughout from due S. at 12 knots. What is the latitude of arrival point?

7. The R.F. of a map is $\frac{1}{126{,}720}$. What is the area in acres of a rectangular aerodrome measuring on the map 0·625 in. by 0·410 in.?

8. The R.F. of a map is $\frac{1}{380{,}160}$. Two points A and B are 1 ft. 9¼ in. apart on the map. At 11.25 a.m. a pilot takes off from A to fly to B and back. His speed on the outward journey is 220 m.p.h. and return speed is 265 m.p.h. At what time does he return to point A, allowing 35 min. for landing and refuelling at B? (Answer to nearest minute.)

9. A twin-engined aeroplane has to fly a round journey (out and back) of 1750 miles at a cruising speed of 180 m.p.h. If each engine consumes 26 gal. of petrol per hour, what weight of petrol must be carried to permit a safety margin of $\frac{1}{10}$th?

10. A bomber starts from an aerodrome at 8.30 p.m. to bomb a target 825 miles away. The outward speed is estimated to be 290 m.p.h. and the return speed 320 m.p.h. Allowing 30 min. for identifying and bombing the target, at what time would the ground staff expect the bomber to return?

11. A motor boat has a fuel capacity of 8 hours while cruising at 10 kt. At 2 p.m. she sets off down stream with a 2 kt. current. At what time must she alter course to return?

12. Which is the faster, and by how many knots: (*a*) 2 N.M. in 8½ min. or (*b*) 7·6 S.M. in 33 min.?

RATIO

A ratio is used to express the relationship between two quantities.

If the cruising speed of an aeroplane is 150 m.p.h. and its maximum speed is 180 m.p.h., the ratio of cruising speed to maximum speed is

$$150:180 \quad \text{or} \quad \tfrac{150}{180}=\tfrac{5}{6}.$$

The ratio therefore is the fraction that cruising speed is of maximum speed, expressed in lowest terms.

It may also be shown as a decimal. Thus

$$\frac{\text{Cruising speed}}{\text{Maximum speed}}=\tfrac{5}{6}=0{\cdot}83.$$

Note that in the following speeds the ratio is in each case the same: $\frac{5}{6}$.

	Cruising speed	Maximum speed
(*a*)	50	60
(*b*)	100	120
(*c*)	85	102

So we see that ratio gives only the relationship between the speeds and is not itself a speed.

If a ratio is given together with an actual value of one of the quantities, then the other quantity can be obtained.

Example. The ratio $\frac{\text{Cruising speed}}{\text{Maximum speed}}$ is $\frac{5}{6}$. The maximum speed is 360 m.p.h. What is the cruising speed?

Cruising speed $=\frac{5}{6}$ maximum speed $=\frac{5}{6}\times 360=300$ m.p.h.

Exercise XXVII.

1. Find the missing quantity in the following:

	Speed ratio $\left(\frac{\text{Cr. sp.}}{\text{Max. sp.}}\right)$	Cruising speed	Maximum speed
(*a*)	$\frac{3}{4}$	—	220 m.p.h.
(*b*)	$\frac{7}{8}$	180 m.p.h.	—
(*c*)	0·81	216 m.p.h.	—
(*d*)	0·78	—	280 m.p.h.

2. In a ratio both quantities involved must be expressed in the same units.

For example, the ratio of a fuel load of 500 gal. and one of 2280 lb. is

$$\frac{(500\times 7{\cdot}2)\text{ lb.}}{2280\text{ lb.}}=\frac{3600}{2280}=\frac{30}{19}.$$

Or, if expressed in gallons:

$$\text{Ratio is } \frac{500\text{ gal.}}{\frac{2280}{7{\cdot}2}\text{ gal.}}=\frac{3600}{2280}=\frac{30}{19}.$$

In the following subtract the reserve from the total fuel load to obtain the normal running load. Then find the ratio $\frac{\text{Normal load}}{\text{Total fuel load}}$.

	Total fuel load	Reserve	Ratio $\frac{\text{Normal load}}{\text{Total fuel load}}$
(a)	2000 lb.	200 lb.	—
(b)	560 gal.	504 lb.	—
(c)	2400 lb.	40 gal.	—

In the following find the reserve:

	Total fuel load	Reserve	Ratio
(d)	2800 lb.	—	$\frac{9}{10}$
(e)	450 gal.	—	$\frac{7}{9}$

3. From the following express the ratio span to length as a fraction and also as a decimal:

	Type	Span	Length
(a)	Spitfire	36 ft. 10 in.	29 ft. 11 in.
(b)	Defiant	39 ft. 4 in.	35 ft. 4 in.
(c)	Junkers 88	59 ft.	46 ft. 6 in.
(d)	Dornier 17	59 ft.	55 ft. 4 in.

4. The "wing loading" of a plane is the ratio of its total weight to the area of its lifting surface, i.e.

$$\text{Wing loading ratio} = \frac{\text{Weight in lb.}}{\text{Sq. ft. lifting surface}}.$$

It is usually expressed as a decimal in lb. per sq. ft.
Find the wing loading ratio in the following:

	Plane	Lifting surface	Weight
(a)	Wright's plane of 1903	600 sq. ft.	925 lb.
(b)	Avro Anson	410 sq. ft.	8,000 lb.
(c)	Wellington Mk I	750 sq. ft.	27,000 lb.
(d)	Lockheed Hudson	551 sq. ft.	17,500 lb.
(e)	Armstrong Whitley	1232 sq. ft.	25,500 lb.
(f)	Dornier 215	592 sq. ft.	18,960 lb.

5. (a) A Bristol Blenheim Mk IV has a wing area of 469 sq. ft. The wing loading ratio is 30·7 lb. per sq. ft. What is the full load weight of the plane?

(b) The full load weight of a Junkers 88 K is 16,980 lb. Find the wing area if the wing loading ratio is 32·9 lb. per sq. ft.

6. In any circle the ratio $\frac{\text{Circumference}}{\text{Diameter}}$ is a fixed quantity, called $\pi = 3{\cdot}14$.

Find the circumferences of circles having the following diameters:

(a) 10 in. (b) 3·5 in. (c) 4·3 cm.

7. A plane at an altitude of 1000 ft. can see a distance of 39 miles in all directions. What is the circumference of this circle of visibility to the nearest mile?

8. Find the diameters of circles with the following circumferences:

(*a*) 2 ft. (*b*) 6·5 cm. (*c*) 2 ft. $10\frac{1}{2}$ in.

9. The circumference of the circle of visibility at an altitude of 2000 ft. is 332·84 miles. Find the visibility distance in any direction.

10. A coastal command plane is searching an area and the radius of search is 3 miles in all directions. Her search circle at the starting point *A* is shown shaded. From *A* the plane follows a semicircular track centred at *B* with 3 miles radius to *C*. Then another semicircular track centred at *A* with 6 miles radius to *D*. Find the distance travelled from *A* through *C* to *D* along the track.

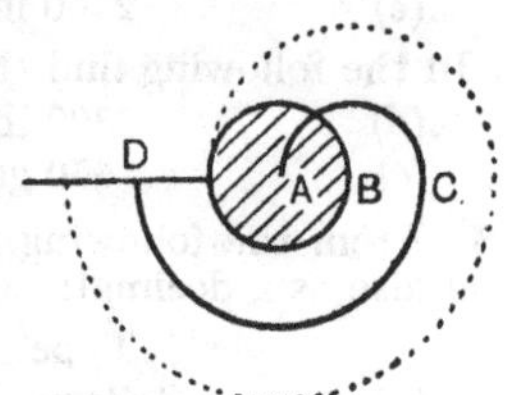

11. In aero-engine design the ratio $\frac{\text{Weight of engine in lb.}}{\text{Horse-power of engine}}$ is a vital factor. This *Weight-Power Ratio*, as it is called, is expressed in lb. per H.P., and the lower the ratio the more useful the engine for aircraft use. The first engine used by the Wright brothers weighed 210 lb. and developed 30 H.P. The weight-power ratio (*W*/*P* ratio) was therefore

$$\frac{\text{Wt. in lb.}}{\text{H.P. developed}} = \frac{210}{30} = 7 \text{ lb. per H.P.}$$

Find the *W*/*P* ratio for each of the following Armstrong Siddeley engines:

	Type	Wt. in lb.	H.P. developed
(*a*)	Tiger	1150	700
(*b*)	Panther	980	560
(*c*)	Jaguar	812	400
(*d*)	Cheetah	710	340

12. Find the H.P. of the following engines:

	Type	Wt. in lb.	*W*/*P* ratio
(*a*)	Bristol Pegasus	1060	1·758 lb. per H.P.
(*b*)	Bristol Mercury	1015	1·676 lb. per H.P.

13. Find the weight of each of the following engines:

	Type	H.P. developed	*W*/*P* ratio
(*a*)	Bristol Pegasus (II)	580	1·827
(*b*)	A.S. Mongoose	155	2·361

14. Speed Ratios. If a machine is driven by a belt connecting two pulley wheels, thus:

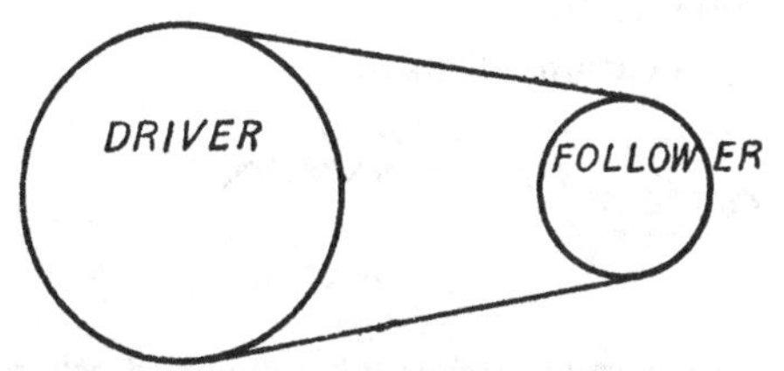

then the speed ratio, i.e.

$$\frac{\text{Speed of follower}}{\text{Speed of driver}} = \frac{\text{Circumference of driver}}{\text{Circumference of follower}}$$

$$= \frac{\pi \times \text{Driver diameter}}{\pi \times \text{Follower diameter}} = \frac{\text{Diameter of driver}}{\text{Diameter of follower}}.$$

If the driver has diameter 2 ft. 6 in. and the follower a diameter of 1 ft., then speed ratio $= \frac{2\frac{1}{2}}{1} = \frac{5}{2}$. When the driver turns at 420 r.p.m., the follower will turn at $420 \times$ speed ratio $= 420 \times \frac{5}{2} = 1050$ r.p.m.

With cogged gearing or cog and chain the speed ratio may be obtained counting the cogs on driver and on follower and using these numbers to find the speed ratio.

From these pulley measurements find the speed ratio and use it to find the speed of the follower:

	Driver diameter	Driver speed	Follower diameter
(*a*)	2 ft. 3 in.	300 r.p.m.	1 ft. 6 in.
(*b*)	2 ft. $7\frac{1}{2}$ in.	240 r.p.m.	1 ft. $1\frac{1}{2}$ in.

Cog and Chain Drive. Find the speed ratio and driver speed:

	Cogs on driver	Cogs on follower	Speed of follower
(*c*)	64	24	300 r.p.m.
(*d*)	32	48	240 r.p.m.

Cogged Gearing. Find the number of cogs on follower:

	Speed of driver	Speed of follower	Cogs on driver
(*e*)	24 r.p.m.	36 r.p.m.	18
(*f*)	240 r.p.m.	210 r.p.m.	28

15. Division of a quantity into parts according to a given ratio.

Suppose 100 ft. of wire is to be divided in the ratio 3 : 5. First, add the numbers in the ratio, i.e. 8, and use this as the denominator of a new fraction. The parts will then be $\frac{3}{8}$ and $\frac{5}{8}$ of the whole, i.e. The two lengths will be $\frac{3}{8}$ of 100 ft. $= 37\frac{1}{2}$ ft. and $\frac{5}{8}$ of 100 ft. $= 62\frac{1}{2}$ ft.

Divide 1 mile of wire into three parts in the ratio (*a*) 3 : 5 : 8; (*b*) 2 : 7 : 13; (*c*) 14 : 15 : 16.

16. On a 480 mile run the ratio of distance covered at cruising speed to that covered at maximum speed was 19 : 5. How many miles were travelled under full throttle?

17. Electrical applications of Ratio.

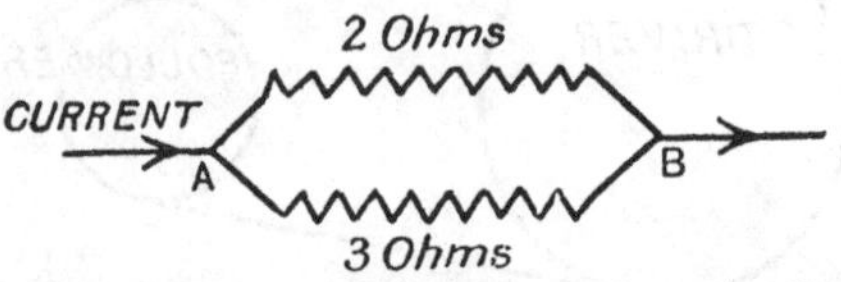

When an electric current arrives at a point *A* where a branching circuit commences, the current divides. The two branch currents reunite at *B* and continue as one current. More current will traverse the smaller resistance (2 ohms) than the larger resistance (3 ohms).

In the above case the ratio of resistances is 2 : 3. But the currents flowing will be in ratio 3 : 2, i.e.

$\frac{3}{5}$ of the total current will be found in the 2 ohm wire
and $\frac{2}{5}$ " " " " 3 ohm wire.

By means of a switch we may alter the resistances of two branches.

If the resistances between the studs are 3, 4, 5 and 6 ohms respectively (see figure) and the switch arm, hinged at *B*, is swung over to stud *A*, then the current will travel to stud *A* and divide.

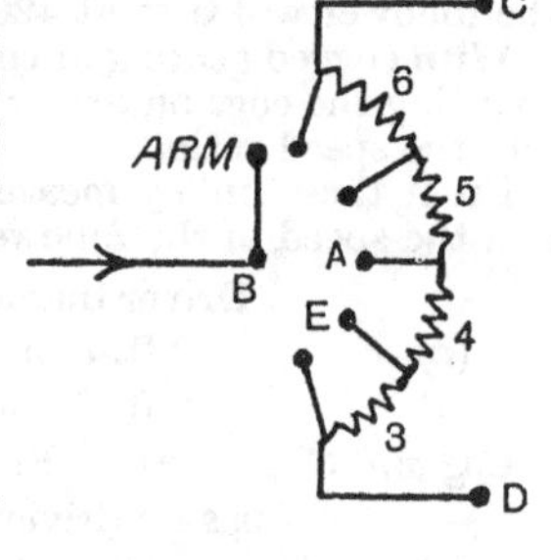

The resistance from *A* to *C*
$=5+6$ ohms (in series) $=11$ ohms.
The resistance from *A* to *D*
$=3+4$ ohms $=7$ ohms.

Thus the resistance ratio is $\frac{AC}{AD}=\frac{11}{7}$ and the current ratio is $\frac{7}{11}$.

What is the resistance ratio, and current ratio, when switch is on stud *E*?

18. Find the resistance ratio in the two branches of the circuit shown, with the switch at *A*, *B* and *C* in turn. (Put branch including *AD* in the numerator.)

What current flows in each branch if the total current is 10 ampères?

19. The Transformer. This instrument consists of a core built up from hollow rectangular sheet-iron stampings. Around it are two windings, the Primary and the Secondary.

The primary winding *PP* (shown thick)

consists of a few turns of thick insulated wire carrying a heavy current, while the secondary winding *SS* contains many turns of fine insulated wire.

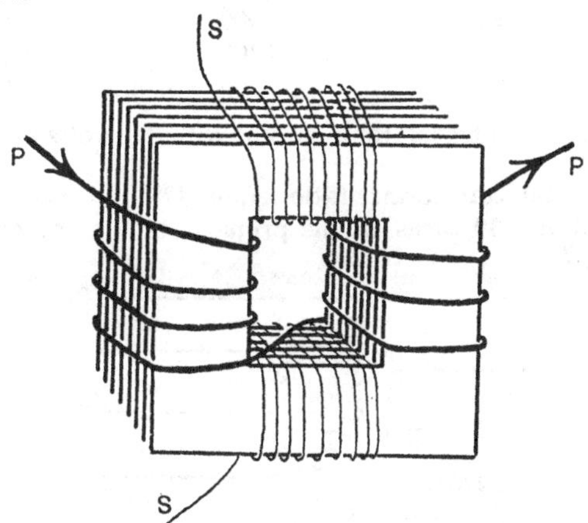

The principle is to "transform" a strong alternating current of several ampères driven by a low voltage into a small current of low ampère strength driven by a high voltage. (This is called a "step-up transformer".) Assuming there are no losses, then the product Volts × Ampères is the same for both coils.

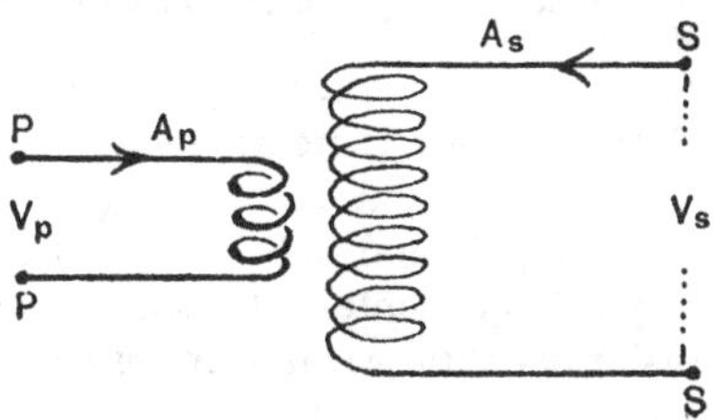

If A_p and V_p represent the ampères and voltage in the primary, and A_s and V_s those in the secondary, then

$$A_p \times V_p = A_s \times V_s.$$

The ratio of the turns in the secondary to the turns in the primary is called the

$$\text{Transformer ratio} = \frac{\text{Number of secondary turns}}{\text{Number of primary turns}}.$$

If there are 5000 secondary turns and 10 primary turns, the transformer ratio is $\frac{5000}{10} = \frac{500}{1}$.

$$\therefore \quad \text{Current in secondary} = A_s = A_p \times \tfrac{1}{500},$$
$$\text{Voltage in secondary} = V_s = V_p \times \tfrac{500}{1}.$$

From the following, calculate (i) the transformer ratio, (ii) secondary voltage:

	Turns of primary	Turns of secondary	Primary voltage
(*a*)	20	8000	15 volts
(*b*)	16	6400	10 ,,
(*c*)	36	1200	12 ,,

20. It is required to transform a current with 220 volts pressure to one of 11,000 volts.

(*a*) What is the transformer ratio required?

(*b*) If there are 32 turns on the primary, how many are needed in the secondary?

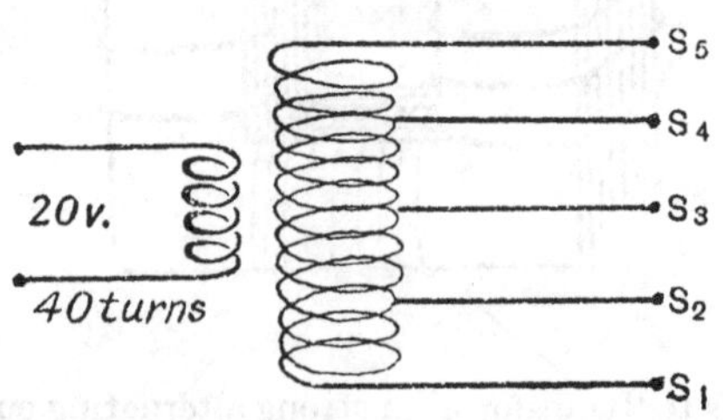

21. In the above secondary we can draw transformed power from terminal S_1 and any of the other four. This will vary the transformer ratio. The total number of secondary turns is 12,000.

Between S_1 and S_2 the secondary voltage has ratio 75/1,
,, S_1 ,, S_3 ,, ,, ,, 150/1,
,, S_1 ,, S_4 ,, ,, ,, 200/1.

How many secondary turns are there between

(*a*) S_1 and S_2? (*b*) S_2 and S_3? (*c*) S_1 and S_3?

(*d*) S_1 and S_4? (*e*) S_4 and S_5? (*f*) S_2 and S_4?

(*g*) What secondary voltage is obtained when drawing from S_3 and S_4?

(*h*) State the transformer ratio when drawing from S_2 and S_4.

PROPORTION

Suppose we construct a scale model of an aeroplane so that the wingspan of the model is 5 ft. while the wingspan of the plane is 60 ft.

The wingspan of the model is thus $\frac{5}{60}$ of the plane's wingspan; and the lengths of all other parts are *in proportion*.

So that

Model's wingspan : Plane's wingspan = Model's length : Plane's length.

If three of the quantities are known, the fourth can be calculated.

Suppose Plane's wingspan = 60 ft.
Plane's length = 72 ft.
Model's wingspan = 5 ft.

Then
$$\frac{\text{Model's length}}{\text{Plane's length}} = \frac{\text{Model's wingspan}}{\text{Plane's wingspan}},$$

i.e.
$$\frac{\text{Model's length}}{72} = \frac{5}{60},$$

$$\text{Model's length} = \frac{72 \times 5}{60} = 6 \text{ ft.}$$

Consider the electrical resistance of a metal wire. We know that as we increase the length of wire we increase the resistance, and the amount of increase in resistance is *proportional* to the increase in length.

So that, in such a case as this, the length of wire and the resistance offered are said to be in *direct proportion*.

EXAMPLE 1. If 500 yd. of wire offer a resistance of 2·8 ohms, what will be the resistance of 135 yd.?

Unit Method.

500 yd. has resistance of 2·8 ohms.

Therefore 1 yd. „ „ $\frac{2\cdot8}{500}$ ohms

and 135 yd. „ „ $\frac{2\cdot8 \times 135}{500}$ ohms $= 0\cdot756$ ohm.

Alternative Method.

The less the length of wire, the less will be the resistance.

Therefore the proportion is $\frac{135}{500}$ of the resistance, i.e.

$$\tfrac{135}{500} \times 2\cdot8 = 0\cdot756 \text{ ohm.}$$

EXAMPLE 2. In 25 min. a plane flies $102\tfrac{1}{2}$ miles. How far will it travel in 1 hour at the same speed?

Unit Method.

In 25 min. plane flies $102\tfrac{1}{2}$ miles.

Therefore in 1 min. „ „ $\frac{102\frac{1}{2}}{25}$ miles

and in 60 min. „ „ $\frac{60 \times 102\frac{1}{2}}{25}$ miles $= 246$ miles.

Alternative Method.

The longer the time, the greater the distance flown.

Therefore the proportion is $\frac{60}{25}$ of the mileage.

$$\text{Distance} = \tfrac{60}{25} \times 102\tfrac{1}{2} = 246 \text{ miles.}$$

EXERCISE XXVIII.

1. A certain metallic wire offers a resistance of 9·8 ohms per mile. Find the resistance of the following lengths:

(*a*) 650 yd. (*b*) 375 yd. (*c*) 1200 yd. (*d*) 337·7 yd.

2. What lengths of this wire would offer the following resistances?
(*a*) 3·12 ohms. (*b*) 0·05 ohm. (*c*) 12 ohms. (*d*) 1·14 ohms.

3. Convert the following into speeds in m.p.h.:

	Time	Distance flown	Speed in m.p.h.
(*a*)	22 min.	83 miles	
(*b*)	1 min. 30 sec.	$2\frac{1}{4}$ miles	
(*c*)	4 min. 30 sec.	$7\frac{1}{2}$ miles	
(*d*)	17 min. 30 sec.	112 miles	

(Answers to first decimal place.)

4. Distance obtained by calculating from the clock time and speed of flight is called "Dead reckoning" (D.R.) distance.
Find the D.R. distances from the following times and speeds:

	Time	Speed	D.R. distance
(*a*)	1 hr. 20 min.	105 m.p.h.	
(*b*)	2 hr. 35 min.	98 m.p.h.	
(*c*)	3 min.	280 m.p.h.	
(*d*)	1 hr. 18 min.	122 knots	
(*e*)	2 hr. 41 min.	432 Km.p.h.	

NOTE. 1 knot is a speed of 1 nautical mile per hour. (All answers in *statute* miles to first decimal place.)

5. On a photograph taken from the air at an altitude of 5000 ft. two objects appear 3 in. apart. If the photographic plate is 6 in. behind the lens, find the actual distance apart of the objects on the ground.

NOTE. $\frac{\text{Distance from lens to plate}}{\text{Distance from lens to ground}} = \frac{\text{Photographic distance}}{\text{True distance on ground}}$.

6. If a plane consumes 30 gal. of petrol in $1\frac{1}{4}$ hr., what will it use in $2\frac{3}{4}$ hr. at the same speed?

7. A plane travels 150 miles on 24 gal. of fuel. What is the range of the plane at the same speed if it carries 84 gal. of unreserved fuel?

8. Estimate the duration of flight, from the following readings for speed, in order to reach these objectives:

	Distance flown		Time	Distance to objective
(*a*)	2 miles	in	$1\frac{1}{2}$ min.	184 miles
(*b*)	$2\frac{1}{4}$ miles	in	$1\frac{1}{4}$ min.	220 miles
(*c*)	$5\frac{1}{2}$ miles	in	2 min. 10 sec.	310 miles
(*d*)	$3\frac{3}{4}$ miles	in	1 min. 40 sec.	145 miles

Inverse Proportion.

Consider now such cases as these:
The *more* men that work on a job the *less* time taken to do it (or it should be!).

The *greater* the cross-sectional area of a wire the *less* the resistance.
The *greater* the speed of a plane the *less* time it takes for the journey.
Such proportions are called *inverse proportion.*

EXAMPLE. If 50 yd. of a certain wire offer 2·4 ohms resistance, what would be the resistance of 80 yd. of similar wire of twice the cross-sectional area?

Here we have *two* factors to consider.
Firstly, the increased length will increase the resistance.
Secondly, the larger area will decrease the resistance.
Treat them separately.

80 yd. of wire would offer $\frac{80}{50} \times 2{\cdot}4$ ohms.

The new wire has twice the cross-sectional area, therefore the resistance will be halved, i.e.

$$\text{Final resistance} = \frac{8\not{0}}{5\not{0}} \times 2{\cdot}4 \times \frac{1}{2} = \frac{19{\cdot}2}{10} = 1{\cdot}92 \text{ ohms.}$$

EXERCISE XXIX.

1. If 120 yd. of a copper wire offer 1·8 ohms resistance, what will be the resistance of 280 yd. of copper wire of $2\frac{1}{2}$ times the cross-sectional area?

2. One copper wire has 1·2 times the cross-sectional area of another. The thicker wire offers 0·4 ohm resistance for 55 yd. length. What resistance will 100 yd. of the thinner wire offer?

3. When burning 128 tons of fuel per day a ship can stay 15 days at sea without refuelling. What will be her cruising time if 160 tons per day are burnt?

4. A ship's cruising time is 18 days when burning 176 tons of fuel per day. What must be the daily consumption to increase the cruising time by 4 days?

5. Electric current in a circuit is measured in *ampères*. The pressure required to drive the current is measured in *volts*.

The current that flows in a circuit depends upon two factors.

(1) The current is *directly* proportional to the voltage, i.e. if the voltage is doubled, the current is doubled.

(2) The current is *inversely* proportional to the resistance in ohms, i.e. if the resistance is doubled, the current is halved.

EXAMPLE. The current flowing in a circuit is 1·5 amp. What will be the current if the voltage is trebled and the resistance doubled?

Again, work this in two stages.

Trebling the voltage will treble the current, i.e.

$$\text{New current} = (1{\cdot}5 \times 3) \text{ amp.}$$

But doubling the resistance will halve the current. Therefore

$$\text{Final current} = (1{\cdot}5 \times 3 \times \tfrac{1}{2}) \text{ amp.} = 2{\cdot}25 \text{ amp.}$$

A current of 1·4 amp. flows in a circuit. What will be the current flowing if the resistance is halved and the voltage doubled?

6. A current of 1·4 amp. flows in a circuit. The voltage is 2·4 volts. What is the new current if the resistance is halved and the voltage 3 volts?

7. A current of 1·8 amp. is driven by a voltage of 2·5 volts through a circuit of 1·4 ohms resistance. What will be the current when the voltage is 3·75 volts and the resistance 3·4 ohms?

8. What will be the effect on a current of halving the voltage and halving the resistance?

INTERPOLATION

In using tables of various kinds, quick and accurate mental proportion is invaluable.

TABLE I

Speed in m.p.h.	Time in minutes					
	10	20	30	40	50	60
10	1·67	3·33	5	6·67	8·33	10
20	3·33	6·67	10	13·33	16·67	20
30	5	10	15	20	25	30
40	6·67	13·33	20	26·67	33·33	40
50	8·33	16·67	25	33·33	41·67	50
60	10	20	30	40	50	60

The above table shows the distances covered at various speeds during 10 min. intervals. Thus in 40 min. at 40 m.p.h. the distance covered is 26·67 miles.

Single interpolation is the use of simple proportion to estimate the distance run at an intermediate speed between two tabulated values.

Suppose we wish to know the distance run in 40 min. at 32 m.p.h.

The answer must be in the 40 min. column and lies between the 30 m.p.h. and 40 m.p.h. values, i.e. it lies between 20 miles and 26·67 miles.

The difference between these is 6·67 miles. 32 m.p.h. is 2 m.p.h. more than 30 m.p.h. and $\frac{1}{5}$ of the difference between 30 m.p.h. and 40 m.p.h.

Hence by mental proportion 2 m.p.h. is $\frac{1}{5}$ of 6·67 = 1·33 miles.

Therefore the distance covered in 40 min. at 32 m.p.h. = 20 + 1·33 = 21·33 miles.

Similarly, by horizontal interpolation, we can find distance run in 43 min. at 30 m.p.h. = ($\frac{3}{10}$ of 5) + 20 = 21·5 miles.

Double interpolation is the use of simple proportion in two operations, for intermediate values of both time and speed.

EXAMPLE. Find the distance run in 36 min. at 44 m.p.h.
This isolated portion of table I will help the explanation:

m.p.h.	Minutes	
	30	40
40	20	26·67
50	25	33·33

By interpolating horizontally for 36 min. at both 40 m.p.h. and 50 m.p.h. we obtain these values:

m.p.h.	36 min.
40	24
50	30

We can then interpolate vertically for the extra 4 m.p.h. between 40 and 50 m.p.h. This obviously is 0·4 of 6 = 2·4 miles.

The answer, therefore, is 26·4 miles.

Mental interpolation, with training and practice, can be quite rapid.

EXERCISE XXX.

1. From table I, interpolate to find distances run in 20 min. for the following speeds:
   (*a*) 15 m.p.h. (*b*) 34 m.p.h. (*c*) 52 m.p.h. (*d*) 28 m.p.h.

2. From table I, interpolate to find distances run at 40 m.p.h. in the following times:
   (*a*) 18 min. (*b*) 33 min. (*c*) 42 min. (*d*) 55 min.

3. From table I, interpolate to find distances run as follows:
   (*a*) 32 min. at 44 m.p.h. (*b*) 16 min. at 56 m.p.h.
   (*c*) 21 min. at 35 m.p.h. (*d*) 35 min. at 33 m.p.h.
   (*e*) 48 min. at 48 m.p.h. (*f*) 2 hr. 10 min. at 30 m.p.h.
   (*g*) 1 hr. 42 min. at 37 m.p.h.

PERCENTAGES

A percentage is usually expressed by the sign %. Thus 17 % is read as 17 per cent, the words "per cent" meaning "per hundred" or "hundredths", i.e.

$$17\% \text{ is } \tfrac{17}{100}.$$

Any percentage may be changed to a fraction by placing it over the denominator 100, i.e.

$$5\% = \tfrac{5}{100}, \quad 110\% = \tfrac{110}{100}, \quad 12\cdot5\% = \frac{12\cdot5}{100}.$$

A fraction may be expressed as a % by multiplying by 100.

Thus $\frac{4}{5}$ expressed as a % is $(\frac{4}{5} \times 100)\% = 80\%$.

Any decimal may be changed to % by multiplying by 100, i.e. by moving the decimal point two places to the right.

Thus $0\cdot25 = 25\%$ and $3\cdot2 = 320\%$.

EXERCISE XXXI.

1. Express these percentages as fractions in their lowest terms:
   (*a*) 15 %. (*b*) 72 %. (*c*) $6\frac{1}{4}$ %.
   (*d*) $102\frac{1}{2}$ %. (*e*) $3\frac{3}{4}$ %. (*f*) 100 %.

2. Convert these percentages to decimal fractions:
   (*a*) 16 %. (*b*) $2\frac{1}{2}$ %. (*c*) 125 %.
   (*d*) $3\frac{3}{4}$ %. (*e*) 14·2 %. (*f*) 1·05 %.

3. Convert these decimal fractions to percentages:
   (*a*) 0·18. (*b*) 2·1. (*c*) 0·34. (*d*) 0·7. (*e*) 0·065. (*f*) 0·084.

4. Express these vulgar fractions as percentages:
   (*a*) $\frac{1}{2}$. (*b*) $\frac{1}{4}$. (*c*) $\frac{3}{4}$. (*d*) $\frac{1}{8}$. (*e*) $\frac{7}{8}$.

5. Express the following as % to one decimal place:
   (*a*) $\frac{1}{3}$. (*b*) $\frac{5}{6}$. (*c*) $\frac{2}{9}$. (*d*) $\frac{3}{13}$. (*e*) $\frac{4}{51}$. (*f*) $\frac{31}{47}$.

6. Find the new speed in the following cases:

	Original speed	% alteration
(*a*)	180 m.p.h.	10 % increase
(*b*)	172 m.p.h.	5 % decrease
(*c*)	92 knots	8 % increase
(*d*)	212 Km.p.h.	17 % increase
(*e*)	208 knots	11 % decrease

7. Find the % change of speed in the following and express as increase or decrease:

	First speed	Second speed
(*a*)	123·5 m.p.h.	108 m.p.h.
(*b*)	115·2 knots	131 knots
(*c*)	226·8 Km.p.h.	198 Km.p.h.
(*d*)	120·8 m.p.h.	100 m.p.h.

8. A pilot flying at 4000 ft. altitude has a visibility of 63·2 miles in all directions. He increases altitude to 5300 ft. and his new visibility is 72·8 miles. Find (*a*) his % increase in altitude, (*b*) his % increase in visibility.

9. From the following details of British Commercial Aeroplanes find (1) cruising speed as % of maximum speed, (2) "pay load" as % of total weight:

	Type	Total weight	Pay load	Maximum speed	Cruising speed
(*a*)	A.W. "Argosy"	18,500 lb.	4000 lb.	109 m.p.h.	95 m.p.h.
(*b*)	H.P. "Heracles"	29,500 lb.	8500 lb.	127 m.p.h.	106 m.p.h.
(*c*)	Short "Scylla"	31,360 lb.	7000 lb.	130 m.p.h.	108 m.p.h.

10. An aeroplane engine specified as 500 rated H.P. at sea level is found to develop only 346·5 H.P. at 10,000 ft. altitude. Find the % fall in power at this altitude (due to the effect of rarefied atmosphere on carburation).

11. The instrument employed in an aeroplane to show the speed in m.p.h. is the Air Speed Indicator. This instrument works by air pressure and is calibrated, i.e. has its dial marks fixed, at sea level. Because the atmosphere is less dense as altitude increases the instrument reads less than the true air speed at altitudes above sea level. Consequently a correction must always be added to the Indicated Air Speed (I.A.S.) to allow for the effect of diminished air density on the instrument.

This correction is 1·75 % of the I.A.S. for every 1000 ft. rise in altitude.

EXAMPLE 1. The indicated air speed for a plane flying at 4000 ft. is 100 m.p.h.

The correction is

$$\left(\frac{\text{Altitude}}{1000}\times 1{\cdot}75\right)\%\text{ to be added} = (4\times 1{\cdot}75)\%\text{ of }100$$

$$= 7\%\text{ of 100 to be added to the I.A.S.}$$

Therefore True air speed = 107 m.p.h.

EXAMPLE 2. Indicated air speed = 140 m.p.h. at altitude 10,000 ft.

Correction is

$$(10\times 1{\cdot}75)\%\text{ of 140 m.p.h.} = 17{\cdot}5\%\text{ of 140 m.p.h.}$$
$$= 24{\cdot}5\text{ m.p.h.}$$

Therefore True air speed = 140 + 24·5 = 164·5 m.p.h.

From the following altitudes and indicated air speeds, obtain the true air speeds:

	Altitude in feet	Indicated air speed
(*a*)	5,000	120 m.p.h.
(*b*)	10,000	120 m.p.h.
(*c*)	7,500	132 m.p.h.
(*d*)	9,150	151·4 m.p.h.
(*e*)	8,270	96·8 m.p.h.
(*f*)	1,100	111·2 m.p.h.

Altitude correction must always be applied to indicated air speeds or otherwise the dead reckoning distance would be greatly in error.

12. A plane is flying at 150 m.p.h. and consuming 30 gal. of fuel per hour. It is found that by reducing speed by 10 % the fuel consumption is reduced by $12\frac{1}{2}$ %. If the tank capacity is 200 gal., of which 10 % must be kept as reserve, find by what % her original range is either increased or decreased by reducing speed.

13. What % error would be made if a pilot plotted a dead reckoning run of 30 nautical miles as statute miles on his chart?

14. What % error would have been made if he plotted 30 statute miles as nautical miles?

15. For the sake of flexibility many electric wires are made up of strands twisted together like strands of a rope. This twisting involves shortening and 3 % is usually allowed for this shortening. Thus in 100 yd. of insulated stranded flex wire, each strand is 103 yd. when straightened out.

Find the actual length of current path in the following lengths of stranded wire:

(*a*) 56 yd. (*b*) 187 yd. (*c*) half a mile. (*d*) 79 kilometres.

16. A lump of duralumin (weldable aluminium alloy) weighing 3 lb. 12 oz. is found on analysis to consist of the following:

Aluminium	3 lb. 8·82 oz.	Manganese	0·36 oz.
Copper	2·52 oz.	Magnesium	0·3 oz.

What is the percentage composition of this alloy? (Answer to nearest first decimal place.)

17. What is the percentage composition of *Y* alloy if analysis of 12·8 grams was as follows?

Aluminium	11·85 gm.	Nickel	0·27 gm.
Copper	0·5 gm.	Magnesium	0·18 gm.

(Answer to nearest first decimal place.)

18. In a descending aircraft the altimeter is lagging by 250 ft. What is its percentage error when recording 8500 ft.?

19. An altimeter is consistently high by 300 ft. at all altitudes. At what recorded altitude will its error be 2 % of the true altitude?

20. An axle of diameter 3 in. is required to fit a bearing so as to have a clearance of 0·1 % of the axle diameter in all directions. What is the diameter of the bearing?

SIGNS AND SYMBOLS

APPLICATION OF THE FOUR RULES

In this subject "symbols", in the form of letters, are used instead of numbers.

It is necessary to understand the methods employed and the rules that must be observed when using letters instead of figures.

In nearly every branch of constructive engineering use is made, for purposes of rapidity or consistency, of a number of standard formulae.

As explained in the section on transposition of formulae it is very often necessary to readjust these formulae so that the value of any one symbol may be expressed in terms of the others.

The following preliminary exercises will help to explain the principles.

Signs and Symbols.

The quantity $3a$ really means $3 \times a$ or simply 3 a's.

$5x$ means $5 \times x$ or 5 x's.

Thus $a+a+a=3a$ and $x+x+x+x+x=5x$.

The signs + and − are used as in arithmetic, and if a quantity has no sign before it (as in $3a$ for example) it is understood to be $+3a$, just as 6 means $+6$ in arithmetic.

The sign always refers to the quantity which follows it.

We may add or subtract in algebra just as in arithmetic.

$$6a+3a=9a \quad \text{or} \quad 4x-x=3x.$$

Remember that the symbols we add or subtract must be the same. It is impossible in arithmetic to take three gallons of petrol from four machine guns!

The + and − signs in algebra mean not only add and subtract, they mean go upward or downward as well.

Imagine a flight of stairs with a landing half way up. The stairs above the landing are the "up", or +, stairs and those below the landing are the "down", or −, stairs.

Thus, if we think of the stairs as a symbol (x), $8x-10x$ means go up 8 stairs and down 10. The answer is $-2x$, or two below the landing.

Different symbols and figures without symbols must be kept separate when adding or subtracting.

EXAMPLE. Simplify:

$$\begin{aligned}5q+3p-2-4q-7p+4 &= 5q-4q+3p-7p-2+4\\ &= q-4p+2.\end{aligned}$$

EXERCISE XXXII.

1. Simplify:

(a) $7b-2b$. (b) $3x-5x$. (c) $13y-2y+y$.
(d) $2z+3z-6z+4z$. (e) $4p-3p-7p$.

2. Simplify the following and find the value of each answer from the value of the symbol:

(a) $3n-6n+4n+2n$ $(n=2)$. (b) $4b+3b-10b$ $(b=\frac{1}{2})$.
(c) $15x-8x+2x-9x$ $(x=3)$.

3. Simplify:

(a) $3s+2t-5s-3t+4+7s$. (b) $12m-2n+4-3m+6n+2$.
(c) $6w-10v-2v-w+5$.

4. Simplify and find the value of:

(a) $5w-6s+6$ (when $w=2$, $s=1$).
(b) $3k+2p+4k-5p$ (when $k=1$, $p=\frac{1}{2}$).
(c) $5x-2y+4x-3y+5$ (when $x=1\frac{1}{2}$, $y=2\frac{1}{2}$).
(d) $4p-\frac{3}{2}q+5p+\frac{1}{2}q$ (when $p=1$, $q=2$).
(e) $\frac{5}{2}a+\frac{b}{3}-\frac{a}{4}+\frac{b}{2}$ (when $a=4$, $b=6$).

Sometimes a symbol may be given a negative value.
Thus if $a=-2$,

$$3a=3(-2)=-6.$$

When a man has no money, but owes £2, he has less than nothing, i.e. $-$£2.

If he owes three such debts he has $3\times(-£2)=-£6$.

If we pay two of the man's debts (i.e. *take away* two of his debts) it is the same as giving the man £4, i.e. $+$£4.

So that $$-2\times(-£2)=+£4.$$

Thus we see that when a $-$ quantity is multiplied by a $+$ quantity the answer is $-$, i.e. $-2\times+3=-6$, or $+2\times-3=-6$, and when two $-$ quantities (or two $+$ of course) are multiplied together the answer is $+$.

This rule is usually remembered thus:

Like signs give $+$,
Unlike signs give $-$.

5. Simplify and evaluate the following:

(a) $3a-5b+2a-2b$ (when $a=2$, $b=-2$).
(b) $6x+2y+4-x-5y$ (when $x=1$, $y=-1$).
(c) $-5x-y-3x-2y$ (when $x=\frac{1}{2}$, $y=-\frac{1}{2}$).

(*d*) $2p-4q+q-3p+3q+p$ (when $p=10$, $q=10$).

(*e*) $5y+2z+3+2y+z$ (when $y=0$, $z=0$).

(*f*) $2y+3c-\frac{c}{2}+\frac{y}{3}$ (when $y=3$, $c=-2$).

(*g*) $\frac{x}{2}+\frac{y}{3}+\frac{z}{4}$ (when $x=1$, $y=2$, $z=-3$).

(*h*) $1{\cdot}2x-0{\cdot}8y+0{\cdot}7x-0{\cdot}3y$ (when $x=2$, $y=3$).

Expanding of Terms, Multiplication.

When symbols are written together they are understood to be multiplied.

Thus $3abc$ may be expanded into $3\times a\times b\times c$.

The product

$a\times a$ is not written aa but a^2,

$b\times b\times b$ is written b^3,

$2\times x\times x\times x\times x$ is written $2x^4$,

and $3\times x\times x\times y\times y\times y$ is written $3x^2y^3$.

The small figure is called an "index" and shows how many times the symbol has been multiplied by itself.

6. Expand in full the following:

(*a*) $3ab^2c^3$. (*b*) $5x^3yz^2$. (*c*) $2a^3bc$. (*d*) $5ap^2q^2$.

7. Expand and evaluate the following:

(*a*) $2a^2b^2c$, (*b*) $3ab^2cd$, (*c*) $2a^2bd^2$, (*d*) $5abc^2$,

when $a=2$, $b=1$, $c=\frac{1}{2}$ and $d=-3$.

8. Expand both terms of the following and collect like symbols and re-index:

(*a*) $3xy^2s\times 4x^3yst$. (*b*) $2ab^2c^2\times 4a^2bcd$.

(*c*) $3mn^2x\times m^3y$. (*d*) $acq\times a^2cq^3r$.

If we examine the answers to Question 8 we find, for example, that

$x\times x^3$ gives x^4 in the answer, i.e. $x^{(1+3)}$,

$q^2\times q^3$ gives q^5 in the answer, i.e. $q^{(2+3)}$.

If we remember, when multiplying, to *add* the indexes (or indices) of the same symbol we can write down the answers without expanding.

9. Work these without expanding:

(*a*) $4bx^2r^2\times rbx^2$. (*b*) $2c^3d^3e^3\times 3ce^2$. (*c*) $stv^3\times stv^3$. (*d*) $(a^2b^2c)^2$.

Addition.

EXAMPLE. Add the following expressions:

$$a-2b+3c; \quad 2c-3a+b; \quad 2b+4a+3c.$$

First write these expressions, so that all like terms are under one another, remembering the signs.

Thus
$$\begin{array}{r} a-2b+3c \\ -3a+\ b+2c \\ +4a+2b+3c \\ \hline 2a+\ b+8c \end{array}$$
By addition

EXERCISE XXXIII.

1. Find the sum of:

(a) $2a-b-c$; $a+2b-c$; $a-b-2c$.

(b) $2x-3y+z$; $3x-y+z$; $4x-3y+2z$.

(c) $2m^2+3an+p^2$; $m^2+4an+2p$; $3m^2+p^2$; $2an+3p$.

(d) $x^2y+2x^2z+y^2z$; $3y^2z-2x^2y+x^2z$; $-x^2z+2y^2z+x^2y$.

(e) $a^3y+2a^2y^2+ay^3$; $5a^2y^2-2ay^3+a^3y$; $3ay^3+2a^3y+a^2y^2$.

(f) $4p^2q+p^3+2pq^2$; $2q^2p+3p^3-2qp^2$; $p^3-pq^2+qp^2$.

(g) $\frac{p}{4}-\frac{q}{3}+\frac{r}{2}$; $\frac{r}{3}+\frac{p}{3}+\frac{q}{3}$; $\frac{2p}{3}-\frac{5r}{6}$.

(h) $1{\cdot}1ax+1{\cdot}7ax^2$; $0{\cdot}7a^2x+0{\cdot}3ax$; $1{\cdot}4ax^2-1{\cdot}6a^2x$.

Subtraction.

EXAMPLE. From $15a-8b+6c$ take $3c-2b+3a$.

First copy the sum with like terms under one another, as before:

$$\begin{array}{r} 15a-8b+6c \\ 3a-2b+3c \\ \hline 12a-6b+3c \end{array}$$

To avoid trouble with minus signs in the line to be taken away, this simple rule is employed: "Change all signs in the bottom line, and add."

2.

	From	Take
(a)	$12b-2c+4d$,	$3d-3c+5b$.
(b)	$2a^2+3ab+3b^2$,	$5ab-2b^2+a^2$.
(c)	$3x^2y-5xy^2$,	$3xy^2-5x^2y$.
(d)	$3p^2qr+4pq^2r+5pqr^2$,	$2pq^2r+5pqr^2-p^2qr$.
(e)	$2x^4+3x^3-5x^2+x+2$,	$3x^3-2x^2+x^4-2x-3$.
(f)	$2a^2+ab+b^2$,	$2a^2-ab+b^2$.
(g)	$3a^2$,	$2ab+b^2$.
(h)	1,	$ab+bc-ac+2$.

Division.

EXAMPLE. Simplify $6a^3b^2c^4 \div 2a^2bc^2$.

This may be written $\frac{6a^3b^2c^4}{2a^2bc^2}$, and expanding,

$$\frac{\overset{3}{\cancel{6}} \times \cancel{a} \times \cancel{a} \times a \times \cancel{b} \times b \times \cancel{c} \times \cancel{c} \times c \times c}{\cancel{2} \times \cancel{a} \times \cancel{a} \times \cancel{b} \times \cancel{c} \times \cancel{c}} = 3 \times a \times b \times c \times c = 3abc^2.$$

3. Expand the following and cancel:

(*a*) $\frac{a^2b^2c^3}{abc}$. (*b*) $\frac{p^3q^2r^3}{p^2qr}$. (*c*) $\frac{8abd^5}{4bd^3}$. (*d*) $\frac{2a^2r^3d}{4rd}$.

(*e*) $\frac{2\pi a^2r^2}{2\pi ar}$. (*f*) $\frac{3\pi^2r^3x^2}{2\pi r}$. (*g*) $\frac{a^4r^5}{a^2r^3}$. (*h*) $\frac{(ar^2)^2}{ar}$.

If we expand $\frac{x^4}{x^2}$ we get $\frac{x \times x \times x \times x}{x \times x} = x^2$.

This may also be obtained by *subtracting* the index numbers

$$\frac{x^4}{x^2} = x^{(4-2)} = x^2.$$

Also
$$\frac{x^5a^6}{x^2a^2} = x^{(5-2)}a^{(6-2)} = x^3a^4.$$

4. Solve these divisions without expanding:

(*a*) $\frac{6a^3b^4c^3}{2ab^2c}$. (*b*) $\frac{8a^5b^4c}{2a^2b}$. (*c*) $\frac{8x^3y^3z^3}{4xz^2}$. (*d*) $\frac{\pi a^3p^5z^2}{ap^3z^2}$.

(*e*) $\frac{3a^2px^2}{2ap^2}$. (*f*) $\frac{(2xyz)^3}{(2xy)^2}$. (*g*) $\frac{4a^4}{2a^3p^2}$. (*h*) $\frac{a^2b^2}{a^2b^2c^2}$.

Brackets.

When quantities are placed in brackets any operation to be performed on the bracket applies to every quantity inside the bracket.

Thus $4(2a+3b)$ means that both $2a$ and $3b$ must be multiplied by 4.

Therefore $4(2a+3b) = 8a+12b$.

Similarly, $-2(x+2y) = -2x-4y$

and $-3(p-2q) = -3p+6q$.

NOTE that $2-(x+2y)$ means that both x and $2y$ are to be taken from 2. The answer is $2-x-2y$.

5. Remove the brackets from:

(*a*) $3(x+y)$. (*b*) $4(2x-a)$. (*c*) $6\left(\frac{x}{2}+\frac{y}{3}\right)$.

(*d*) $3(a+b+c)$. (*e*) $-4(x-y)$. (*f*) $-\frac{5}{2}(2a+4b-c)$.

(*g*) $3(ax+by)$. (*h*) $3a(ax+by)$.

6. Write in bracket form (i.e. place outside a bracket all that is common to all terms):

(a) $2x+2y$. (b) $3a+3b+6c$. (c) $4x^2+4y^2$.
(d) $-3ax+3bx$. (e) $ax+ay+az$. (f) $bx^2+2bxy+by^2$.
(g) $3px^2+6py^2$. (h) $-3a^2r-6ar^2$.

Check each one mentally by removing the bracket again.

When two brackets are side by side, thus:

$$(x+y)(a+y),$$

it means that both quantities in the first bracket must be multiplied by both quantities in the second, i.e.

$$\begin{aligned}(x+y)(a+y) &= x(a+y)+y(a+y)\\ &= ax+xy+ay+y^2.\end{aligned}$$

Example. $(a-z)(a+z)$ means

$$\begin{aligned}a(a+z)-z(a+z) &= a^2+az-az-z^2\\ &= a^2-z^2.\end{aligned}$$

7. Work out the following:

(a) $(x-y)(x+y)$. (b) $(a+b)(a+b)$.
(c) $(p+q)^2$. (d) $(a^2-x^2)(a^2+x^2)$.
(e) $(a+x)(a^2-x^2)$. (f) $(2p+q+r)(p+2q-r)$.
(g) $3^2(a+x)^2$. (h) $\frac{1}{2}\left(\frac{a}{2}-\frac{x}{2}\right)\left(\frac{a}{2}+\frac{x}{2}\right)$.

Factorising.

We know that $3(x+y)=3x+3y$.

So that 3 and $(x+y)$ are said to be factors of $3x+3y$, because these two quantities when multiplied together produce $3x+3y$.

Similarly, we have seen that

$$\begin{aligned}(x-y)(x+y) &= x(x+y)-y(x+y)\\ &= x^2-y^2.\end{aligned}$$

So that the factors of x^2-y^2 are $(x-y)$ and $(x+y)$.

Any two or more quantities which when multiplied together produce a given quantity are factors of that quantity:

$$\begin{aligned}(x+1)(x-3) &= x(x-3)+1(x-3)\\ &= x^2-3x+x-3\\ &= x^2-2x-3.\end{aligned}$$

So that the factors of x^2-2x-3 are $(x+1)$ and $(x-3)$.

In the expression $az+bz$ there are two terms and z appears in both as a multiplier.

Hence $az+bz=z(a+b)$.

Similarly, $a(x+y)+b(x+y)$ contains two terms and in each there is a common factor $(x+y)$.

Therefore $a(x+y)+b(x+y)=(a+b)(x+y)$.

8. Factorise:

(a) $a(b+c)+c(b+c)$. (b) $x(y-z)+y(y-z)$.
(c) $p(2q+r)-2q(2q+r)$. (d) $a^2(3p+q)+b^2(3p+q)$.
(e) $ax+ay+bx+by$. (f) $cx+dy+cy+dx$.
(g) $ax+x^2+ay+xy$. (h) $p^2+pq+pq+q^2$.

Suppose we now expand $(2x+3)^2$, i.e.

$$\begin{aligned}(2x+3)(2x+3) &= 2x(2x+3)+3(2x+3)\\ &= 4x^2+6x+6x+9\\ &= 4x^2+12x+9.\end{aligned}$$

Thus the factors of $4x^2+12x+9$ are equal and both are $(2x+3)$.

Suppose we had been asked to factorise $4x^2+12x+9$ into two bracketed quantities.

In two brackets there are four positions, 1, 2, 3 and 4, as shown:

$$(1 \quad 3)(2 \quad 4)$$

Whatever is placed in positions 1 and 2 must, when multiplied together, produce $4x^2$.

And whatever is placed in 3 and 4 must produce 9.

The middle term $12x$, in the problem, is obtained from the sum of:

The product of 3 2—called "product of means" and product of 1 4—called "product of extremes", i.e.

$$3 \quad 2+1 \quad 4=12x.$$

The factors are $(2x+3)(2x+3)$.

EXAMPLE. Factorise $x^2+13x+30$.

First step: $(x \quad)(x \quad)$, giving x^2.

Second step: $(x \quad 1)(x \quad 30)$ or $(x \quad 2)(x \quad 15)$ or $(x \quad 3)(x \quad 10)$ or $(x \quad 5)(x \quad 6)$, each giving 30.

Third step: from means and extremes, the only additions giving $+13x$ are $3x$ and $10x$, or $-2x$ and $15x$. But the last term in the original expression is $+30$ as that $-2x$ and $15x$ are inadmissible.

The factors therefore are $(x+3)(x+10)$.

Check these examples by positional multiplication, i.e. (1.2); (3.4); means and extremes:

Expression	Factors
x^2+4x+4.	$(x+2)(x+2)$.
x^2-4x+4.	$(x-2)(x-2)$.
x^2+5x+4.	$(x+4)(x+1)$.
x^2-3x-4.	$(x-4)(x+1)$.
x^2-5x+4.	$(x-4)(x-1)$.
$2x^2-x-6$.	$(2x+3)(x-2)$.

TRANSPOSITION OF FORMULAE

A formula is an easily remembered symbolical expression showing some practical relationship.

EXAMPLE 1. In the simple formula

$$C=\frac{E}{R}\text{ (Ohm's law)}$$

used in an electrical circuit,

C is the current in Ampères,
R is the resistance in Ohms,
E is the electrical pressure in Volts.

By "transposition", any desired symbol in a formula can be isolated to simplify a solution.

The above formula may just as readily be written

$$E=CR \quad\text{or}\quad R=\frac{E}{C}$$

if we wish to find E or R respectively.

EXAMPLE 2. Examine this formula:

$$R_t=R_0(1+\alpha t).$$

This is used to find R_t, i.e. the resistance of a wire at t° Centigrade, when R_0, the resistance at 0° C., and α, the temperature coefficient of the metal, are known.

Suppose R_t and R_0 are known and we wish to find t.

Then $$R_t=R_0(1+\alpha t)=R_0+R_0\alpha t,$$

i.e. $$R_t-R_0=R_0\alpha t.$$

$$\therefore\ t=\frac{R_t-R_0}{R_0\alpha}.$$

EXAMPLE 3. From the formula $\frac{R_1}{R_2}=\frac{a+kd}{a-kd}$ find a value for d.

By cross multiplication

$$R_1(a-kd)=R_2(a+kd),$$

i.e. $$R_1a-R_1kd=R_2a+R_2kd,$$

$$-R_1kd-R_2kd=R_2a-R_1a,$$

and $$R_1kd+R_2kd=R_1a-R_2a.$$

Therefore $$dk(R_1+R_2)=a(R_1-R_2),$$

and $$d=\frac{a(R_1-R_2)}{k(R_1+R_2)}.$$

EXERCISE XXXIV.

In Questions 1–4 express the required letter in terms of the others.

1. (a) $C=2\pi r$, find r. (b) $C^2=\frac{f}{12}$, find f.

(c) $t^2=\frac{\pi^2 l}{8}$, find l. (d) $H=\frac{2}{3}nd^2$, find n.

(e) $s=u+ft$, find f. (f) $s=ut+\frac{1}{2}ft^2$, find f.

2. (a) $V=v(1+at)$, find a. (b) $S=c+\frac{cp}{100}$, find p.

(c) $S=2\pi r^2+2\pi rh$, find h. (d) $S=\frac{n}{2}\{2a+(n-1)d\}$, find a.

(e) $S=\frac{W(v-u)}{2g}$, find u. (f) $P=\frac{2\pi(l+3h^2)}{3gh}$, find l.

(g) $S=\frac{AKN}{36\times 10^5\pi d}$, find d. (h) $f^2=\frac{1}{4\pi^2LC}$, find L.

3. (a) $\frac{2}{r}=\frac{1}{v}+\frac{1}{u}$, find v. (b) $C=\frac{nE}{nr+R}$, find n.

(c) $\frac{m}{n}=\frac{ax+b}{cx+d}$, find x. (d) $c=\frac{ax-2by}{ax+2by}$, find x.

(e) $\frac{R_1}{R_2}=\frac{a+\mu d}{a-\mu d}$, find μ. (f) $x=y\left(\frac{2t}{p+q}-1\right)$, find t.

4. (a) $RT=\left(P+\frac{a}{V^2}\right)(V-b)$, find P.

(b) $V=\pi h^2\left(r-\frac{h}{3}\right)$, find r.

(c) $\frac{V^2}{v^2}=\frac{3(M+2m)}{2(M+3m)}$, find m.

(d) $S=\pi g(l+p)+\pi f$, find l.

(e) $k^2=R^2+2\pi l(k+R)$, find l.

(f) $\frac{1}{p^2}=\frac{1}{a}+\frac{1}{b}$, find a.

(g) $t^2=\frac{4\pi^2(2Wa^2+Mb^2)}{Mgb}$, find M.

5. (a) $V=\pi l\{R^2-(R-t)^2\}$, find R in terms of the other symbols and the value of R when $V=20$, $l=14$, $\pi=\frac{22}{7}$, $t=2$.

(b) $\frac{m-p}{m}=\frac{1+xs}{1+xt}$, find x in terms of other symbols and its value when $m=10$, $p=8$, $t=20$, $s=-4$.

(c) If $V=\pi r^2h$ and $S=2\pi r(r+h)$, obtain a value for h which does not contain π.

(d) If $\frac{1}{a}+\frac{1}{b}=\frac{2}{R}$ and $m=\frac{a-R}{R-b}$, find a value for m in terms of a and b, not including R.

(e) If $s=ut+\frac{1}{2}ft^2$ and $v=u+ft$, find v in terms of s, f and t only.

(f) If $2a=4t+3t^2$ and $b=2+3t$, find an expression for a containing b but not t.

GRAPHS

STATISTICAL GRAPHS

Plotting of Points.

It is a common practice to demonstrate statistical details or recorded facts by means of "graphs" (or more correctly "charts"). A true graph will be explained later.

For this purpose it is most convenient to use "squared" paper, ruled in $\frac{1}{10}$ths of an inch or in centimetres and millimetres.

Two lines AB and AC are drawn at right angles and are called axes. Suppose we wish to show a chart of daily sales of a business.

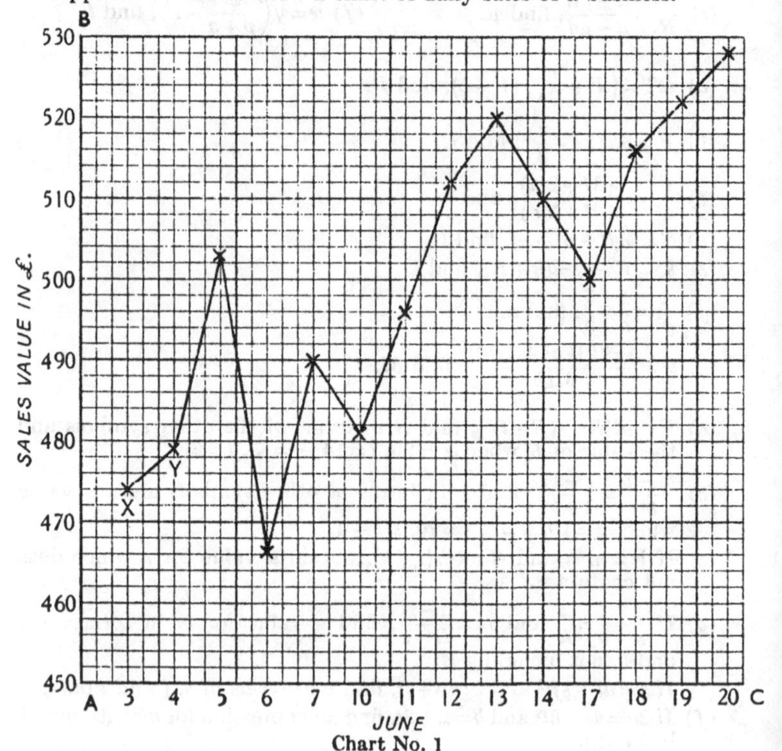

Chart No. 1

The utility of this is to demonstrate quickly, by means of a picture, growth or decline of sales.

In this case the horizontal axis *AC* would be marked off in equal divisions, say three small squares, representing one day.

The vertical scale *AB* would be divided into divisions suitable to represent the amount of the daily sales, allowing a margin for rise or fall (see Chart No. 1).

The details of sales for the first 14 business days in June were as shown in the following table:

June	3	4	5	6	7	10	11
Value of sales in £	474	479	503	466	490	481	496
June	12	13	14	17	18	19	20
Value of sales in £	512	520	510	490	516	522	528

The first point of the chart (£474 sales on June 3rd) is found by travelling horizontally from the 474 mark on the vertical scale to the point immediately above the June 3rd mark on the horizontal scale. This is point *X* on the chart. *Y* is found similarly from the 479 mark to a point immediately above the June 4th point—and so on.

These points are usually joined together by straight lines, which serve as a means to carry the eye from one point to the next.

It must be realised that a chart or "statistical representation" such as this has no value except as a record of fact. It is of no use for estimation or forecasting. It can in fact be most misleading.

For instance, although the sales at 6 p.m. (say) on June 13th were £520 and at 6 p.m. on June 14th £510, it does *not* follow that half-way between these two times (6 a.m. on June 14th) the sales were half-way between £520 and £510. They certainly would not have been.

The principal value of these charts is either for comparison (week to week, or year to year) or as an indication of a general trend. In the case of Chart No. 1 the general *trend* is a rise in sales.

Exercise XXXV.

1. Plot an egg-laying chart for 1000 birds from April 1st to April 15th if these were the number of eggs collected each day: 821; 845; 866; 848; 855; 861; 863; 867; 831; 858; 862; 870; 870; 859.

2. Plot a mileage chart for a reconnaissance plane for 10 consecutive days when the distances flown daily in miles were: 541; 311; 688; 729; 530; 415; 704; 489; 521; 693.

3. Plot a rainfall chart for Kirkwall from details given in Exercise XI, Question 6.

4. Draw a temperature chart for a patient whose temperature is taken at 6 a.m. and 6 p.m. daily for one week with these readings:

Monday		Tuesday		Wednesday		Thursday	
6 a.m.	6 p.m.	6 a.m.	6 p.m.	6 a.m.	6 p.m.	6 a.m.	6 p.m.
101·2	101·4	101·3	102·1	102·1	101·7	101·1	100

Friday		Saturday		Sunday	
6 a.m.	6 p.m.	6 a.m.	6 p.m.	6 a.m.	6 p.m.
100	99·1	99·3	98·6	98·5	98·4

Co-ordinates.

Positions of fixed points are often given by co-ordinates, i.e. by two measurements:

(*a*) Distance N. of a base line;
(*b*) Distance E. of a vertical line.

In Chart No. 2 the position ⊙*A* is 2·1, 2·0 and *B* is 0·755, 4·0.

NOTE. The position of any point that is not on the junction of any two small squares must be made by mental division, and it is possible to be quite accurate to $\frac{1}{10}$th of a small square $= \frac{1}{100}$th of an inch.

5. Name the LETTERS with these co-ordinates:

(*a*) 0·5; 1·0. (*b*) 2·025; 4·35. (*c*) 2·05; 1·25.
(*d*) 0·6; 2·5. (*e*) 2·325; 2·15. (*f*) 0·7; 3·4.

6. Give the co-ordinates of the following letters:

(*a*) *C*. (*b*) *H*. (*c*) *M*. (*d*) *D*. (*e*) *N*. (*f*) *P*.

Mapped Positions. Reporting and Locating.

The horizontal parallels of latitude and the vertical parallel meridians of longitude on a Mercator Chart enable us to use latitude and longitude as co-ordinates of places when describing position.

Examine carefully Chart No. 3.

It extends from the 51° N. latitude parallel at the bottom to the 53° N. latitude parallel at the top and parallels are shown at 10′ intervals.

It also extends West to East from longitude 0 to longitude 2° E. and meridians are again drawn at 10′ intervals.

The left border shows individual minutes of latitude, and the bottom border individual minutes of longitude.

Two dotted "isogonal" lines (lines joining places of equal magnetic variation) are shown.

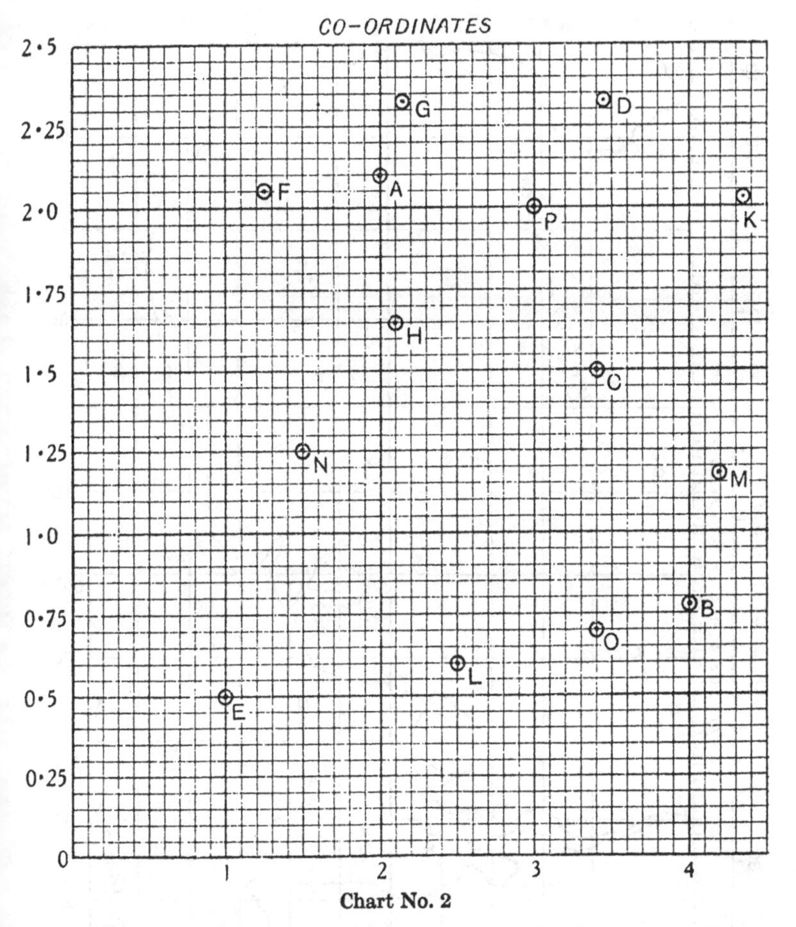

Chart No. 2

In reporting positions from a chart the axes of reference are always the lines of latitude and longitude, of exact degrees, immediately to the S. and W. of the position under report.

These are given code letters. The latitude parallels in this chart might be coded thus:

$$51^\circ \text{ N.} = AB;\ 52^\circ \text{ N.} = CD;\ 53^\circ \text{ N.} = EF,$$

and the meridians thus:

$$0 = UV;\ 1^\circ \text{ E.} = WX;\ 2^\circ \text{ E.} = YZ.$$

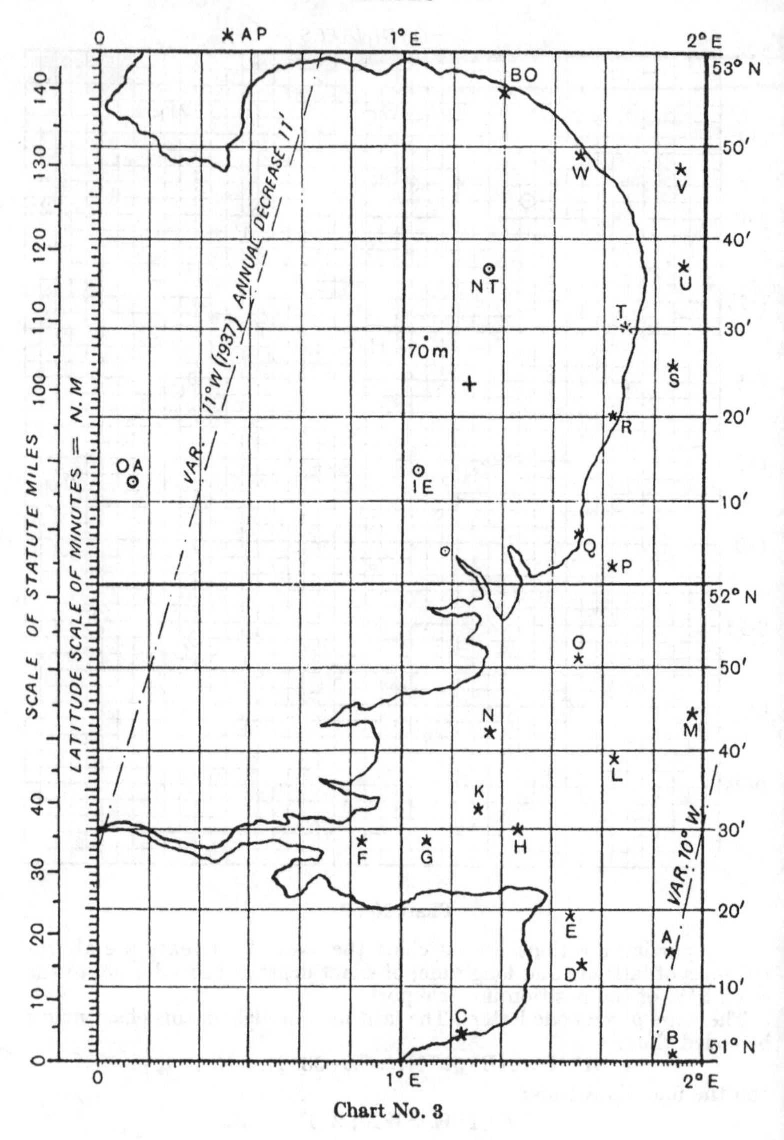
0
1° E
2° E
53° N
50′
40′
30′
20′
10′
52° N
50′
40′
30′
20′
10′
51° N
0
1° E
2° E
0
10
20
30
40
100
110
120
130
140
SCALE OF STATUTE MILES
LATITUDE SCALE OF MINUTES = N.M
VAR. 11° W (1937) ANNUAL DECREASE 11′
VAR. 10° W.
AP
BO
W
V
NT
U
T
70m
S
R
OA
IE
Q
P
O
N
M
L
K
H
F
G
E
A
D
C
B

Chart No. 3

When reporting a position the code letters of latitude precede those of longitude.

Thus *ABWX* denotes that the bases are 51° N. lat. and 1° E. long.

Since the code letters indicate the exact degrees we need only state the number of minutes of latitude that the position is N. of 51° N. and number of minutes of longitude E. of 1° E.

EXAMPLE. Light *B* is *ABWX* 0154. Light *A* is *ABWX* 1453.

7. Report the following lights as to position:

(*a*) *E*. (*b*) *G*. (*c*) *F*. (*d*) *AP*. (*e*) *BO*. (*f*) *R*.

8. Locate the following and give the charted LETTER:

(*a*) *CDWX* 0635. (*b*) *CDUV* 1208. (*c*) *CDWX* 3818.
(*d*) *ABWX* 5135. (*e*) *CDWX* 2653. (*f*) *ABWX* 3942.

CONTINUOUS OR LOCUS GRAPHS

Straight Line Graphs.

These are true graphs, and differ from the statistical charts so far used in one most important point. They *can* be and *are* used to determine intermediate values from known recorded facts.

Suppose we consider the electrical resistance of a length of wire, which, as we have already seen, depends upon the length of the wire.

If 500 yd. of the wire offer 4 ohms resistance and 125 yd. offer 1 ohm, we may readily construct a graph from which the resistances of various odd lengths of this wire may be obtained without calculation (Graph No. 1).

On the axes *AB*, *AC*, two convenient scales are selected to represent the quantities—in this case yards and ohms—that are being dealt with.

The point of intersection of the axes, called the "point of origin", marks the commencement of both scales.

The scale of resistance along *AC* is 10 small squares (or 1 in.) = 1 ohm.

The scale of length, along *AB*, is 10 small squares (1 in.) = 100 yd.

Plot the points *X* and *Y* (500 yd. offer 4 ohms and 125 yd. offer 1 ohm).

Obviously, the resistance of a wire 0 yd. long is 0 ohm. So we expect to find the straight line joining *X* and *Y* to pass through the point of origin (*A*).

Every point on this line *AYX* shows a relationship between "length of wire in yards" and its "resistance in ohms".

What is the resistance of 300 yd. of this wire?

The point *P*, on the 300 yd. line of the distance scale, meets the graph on the 2·4 ohm line of the resistance scale.

Therefore the resistance of 300 yd. of wire is 2·4 ohms.

EXERCISE XXXVI.

1. From the graph find the resistances in ohms of the following lengths of wire:

(*a*) 210 yd. (*b*) 100 yd. (*c*) 360 yd.
(*d*) 235 yd. (*e*) 470 yd. (*f*) 342 yd.

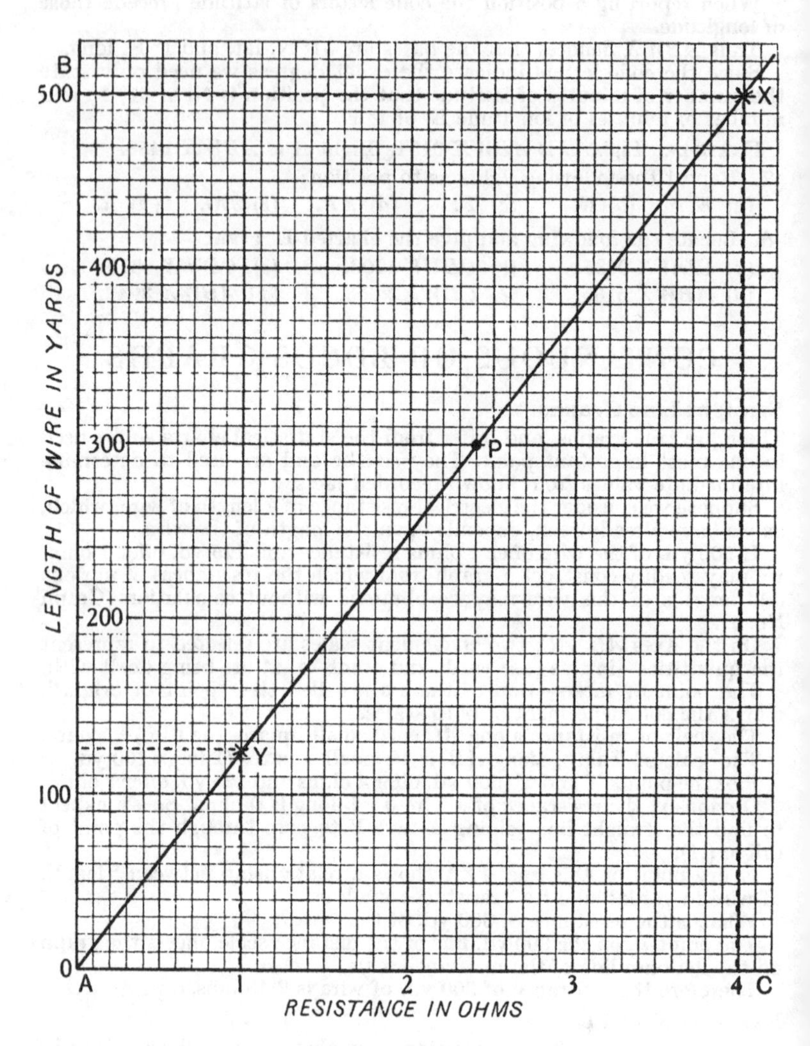
B
500
400
300
200
100
0
A
1
2
3
4
C
X
P
Y
LENGTH OF WIRE IN YARDS
RESISTANCE IN OHMS

Graph No. 1

2. What lengths of wire in yards will offer these resistances?

(*a*) 0·4 ohm. (*b*) 4·4 ohms. (*c*) 2·65 ohms.
(*d*) 2·5 ohms. (*e*) 1·14 ohms. (*f*) 3·34 ohms.

3. Construct a graph to convert kilometres into miles.

Use the vertical scale for kilometres, 1 Km. $= \frac{1}{10}$ in. and the horizontal scale for miles, 1 mile $= \frac{1}{10}$ in.

Plot the two points, 0 Km. = 0 miles and 66 Km. = 41 miles.

From the graph convert

(*a*) 20 miles into kilometres. (*b*) 56 kilometres into miles.

4. Construct a graph to convert miles into nautical miles.

Choose your own scales and plot

0 miles = 0 nautical miles,
76 miles = 66 nautical miles.

From the graph express

(*a*) 19 miles in nautical miles. (*b*) 19 nautical miles in miles.

5. Draw a graph for changing ° C. into ° F., using the following two points:

0° C. = 32° F. and 100° C. = 212° F.

From the graph convert

(*a*) 113° F. into ° C. (*b*) 60° C. into ° F.

6. Draw a graph to show the connection between temperature and altitude from the following:

Sea level (altitude 0 ft.): Temperature 15° C.
Altitude 5000 ft.: Temperature 5·1° C.

From the graph find

(*a*) Temperature at 3500 ft.
(*b*) Altitude at which temperature is 10° C.

7. Given that 300 metres = 984 ft., plot a graph, using horizontal scale 1 in. = 100 metres and vertical scale 1 in. = 200 ft.

From the graph convert the following metres into feet:

(*a*) 225. (*b*) 380. (*c*) 112. (*d*) 168.

Convert the following feet to metres, and check your answers by using 1 metre = 3·28 ft.:

(*e*) 1200. (*f*) 725. (*g*) 416. (*h*) 207.

8. Construct a time and distance graph for use at ground speed 100 m.p.h. to determine distance flown in miles for any interval of time from 0 to 60 min. Use horizontal scale $\frac{1}{2}$ in. = 10 miles; vertical scale 1 in. = 10 min.

From the graph find the time taken (to nearest min.) to fly

(*a*) 33 miles. (*b*) 46 miles. (*c*) 72 miles. (*d*) 21 miles.

And distance flown, to nearest mile, in

(*e*) 42 min. (*f*) 12 min. (*g*) 23 min. (*h*) 55 min.

9. Troops set out on a route march at 9.0 a.m. They march for 50 min. and rest for the last 10 min. in each hour. During each hour they cover 3 miles. At 12.30 p.m. a motor cyclist sets out to overtake them, travelling at a constant speed of 20 miles per hour. Construct a graph to determine the time and how far the troops have travelled when the motor cyclist overtakes them.

Curved Graphs.

All the graphs drawn up to the present have been straight line graphs.

Graphs are by no means all of this type. Many graphs must be drawn by plotting a number of points (from their co-ordinates) and joining them by as natural and continuous a curve as will fit them.

Examine this table:

Time in minutes	Altitude in feet	Rate of climb in feet per min.
0	0	1100
5	5,000	900
11·25	10,000	700
19·55	15,000	500
32·05	20,000	300
47	25,000	100

This table, giving details of a climbing aeroplane, shows (*a*) altitudes reached at different times from the start, i.e. zero; (*b*) the rate at which the plane was climbing at various altitudes.

These facts may be represented by two graphs on the same diagram, using one common horizontal scale of time and two separate vertical scales, as shown in Graph No. 2.

10. The following table shows the distance of visibility in nautical miles at certain heights of eye, in feet, above sea level:

Height of eye in feet	Distance in nautical miles
5	2·57
10	3·63
15	4·45
20	5·13
25	5·74
30	6·29
35	6·79
40	7·26

Draw the graph of these points, when plotted, and from it determine the visibility at:

(*a*) 23 ft. (*b*) 38 ft.

Find also the height of eye when visibility is:

(*c*) 6 N.M. (*d*) 4 N.M.

11. Use Graph No. 2 and find from the "altitude and time curve" how many minutes after the start the plane is at these altitudes:

(*a*) 12,500 ft. (*b*) 18,000 ft. (*c*) 7500 ft.

From the "rate of climb" curve find the climbing rate after the following times:

(*d*) 10 min. (*e*) 20 min. (*f*) 45 min.

Using both curves find the rate of climb at the following altitudes:

(*g*) 12,500 ft. (*h*) 17,500 ft. (*i*) 22,500 ft.

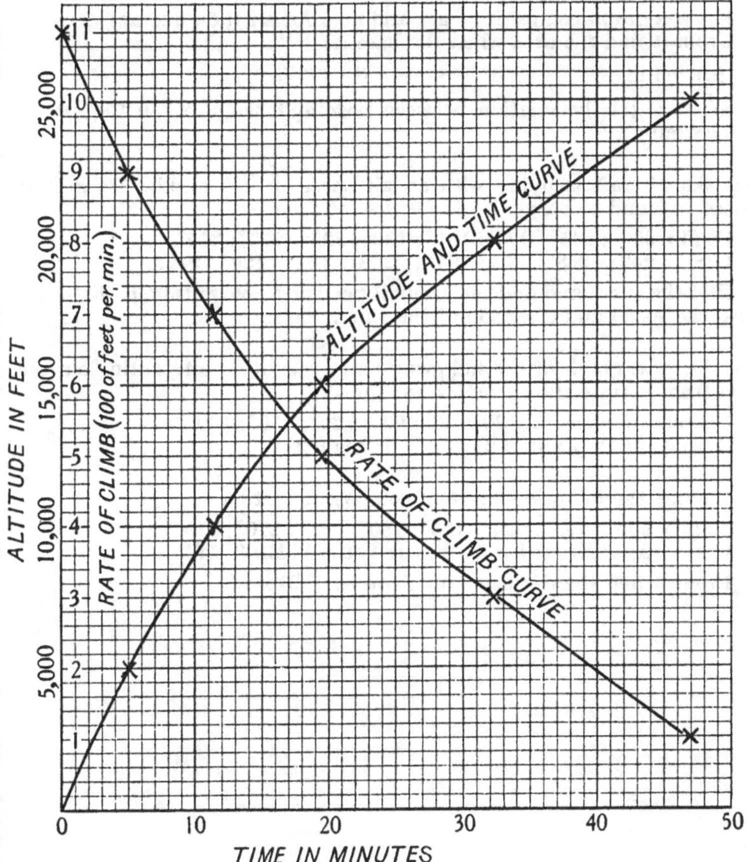

Graph No. 2

12. A piece of wire 24 in. long can be bent to form a rectangle 11 in. high by 1 in. base. Its area would be 11 sq. in. It might be bent to form all the rectangles whose measurements are as follows:

Height in in.	1	2	3	4	5	6	7	8	9	10	11
Base in in.	11	10	9	8	7	6	5	4	3	2	1
Area in sq. in.	11	20	27	32	35	36	35	32	27	20	11

Draw a graph connecting base and area for all these rectangles. From it find the area when the base is:

(*a*) 5·5 in. (*b*) 3·5 in.

What is (*c*) the relation between base and height when the area is a maximum?

13. An aircraft is flying with an air speed of 220 m.p.h. Its ground speed is found to vary with the course as follows:

Co (T) 000°; 045°; 090°; 135°; 180°; 225°; 270°; 315°.
G/S 242; 258; 255; 234; 202; 182; 190; 212.

Plot the curve and find the ground speed on a course 130° T. Estimate also the speed and direction of the wind.

14. Draw a graph of the following half cycle of an alternating current and from it state the current value for 45°:

Degrees of cycle: 0°; 30°; 60°; 90°; 120°; 150°; 180°.
Current in amp.: 0; 6·2; 11; 12·5; 11; 6·2; 0.

15. Draw the following curve of compass deviation:

Co (M) 000°; 045°; 090°; 135°; 180°; 225°; 270°; 315°.
Dev. 0; 4° E.; 1° E.; 3° W.; 0; 4° E.; 1° E.; 3° W.

From it find the deviation on a course of 214° M.

GEOMETRY

Much of the mathematical work in connection with aerial navigation has to be done by drawing with the aid of geometrical instruments.

These include set squares, parallel rulers, compasses and dividers, and the protractor.

Accuracy with the use of any of these instruments is largely a matter of care and practice.

The student should not be disturbed if his answers are not exactly in accordance with the book answers, as small errors are unavoidable and they accumulate during a drawing. The answers, however, will show if any serious error is being made.

It is assumed that the student is conversant with the meaning of "course", "track" and "bearing" (true, magnetic or compass) as these terms are employed in the following chapters.

Remember. In practical navigation there is no answer but your own. Work carefully, and as far as possible balance your errors.

SIMILAR TRIANGLES

Consider fig. 1 and fig. 2.

The two triangles *ABC* have the same length of base *AB* and each base is divided into ten equal parts.

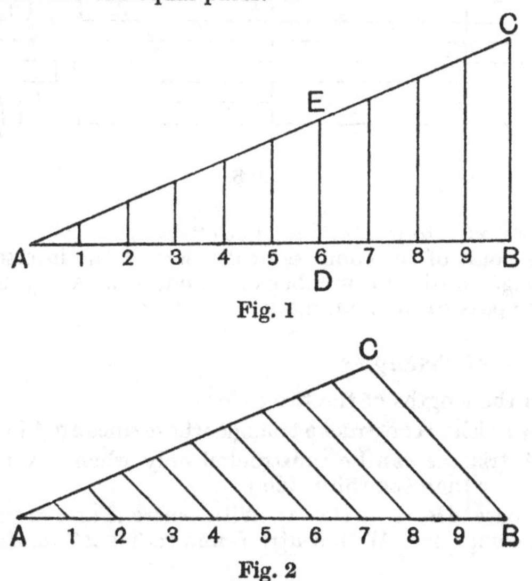

Fig. 1

Fig. 2

The triangle in fig. 1 has a right angle ABC and the one in fig. 2 is not a right-angled triangle. In each of them BC is 1 in. in length.

If, from the numbered points, a set of parallel lines is drawn across the triangles, and each line is parallel to BC, we produce a series of *similar* triangles.

Similar triangles have all their corresponding angles equal, or are, as it is termed, *equiangular.*

In fig. 1 there are now ten triangles and each of the nine new ones is similar to ABC. The same applies to fig. 2.

When triangles are similar their sides are proportional.

This means that since the side AD in the triangle ADE (fig. 1) is $\frac{6}{10}$ of AB, then DE will also be $\frac{6}{10}$ (or 0·6) of BC.

Since BC was 1 in. in length, the perpendiculars from points 1, 2, 3, etc. are respectively 0·1, 0·2, 0·3, etc. in. long.

Although in fig. 2 no right angle was used the lines are nevertheless proportional and starting from A represent 0·1, 0·2, 0·3, etc. in.

This principle of similar triangles is made use of in the *diagonal scale* (fig. 3).

This scale enables us to measure 0·01 in. The diagonal scale is 4 in. long and graduated in inches to the left of O, while the inch to the right of O is marked in $\frac{1}{10}$ths.

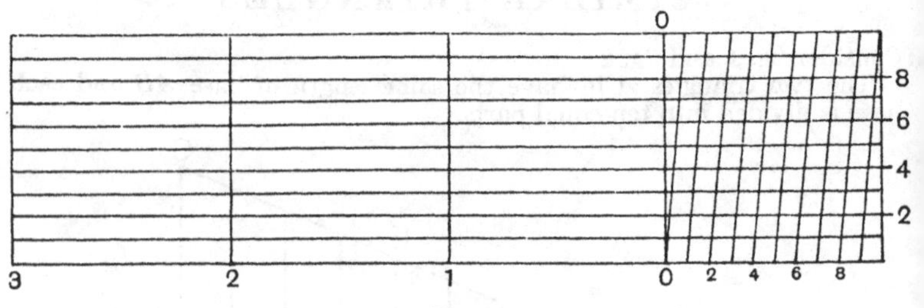

Fig. 3

Suppose we wish to mark off a length 2·64 in.

Place the point of the compass or dividers at the intersection of the number 6 diagonal with the number 4 horizontal and along this horizontal open the compass to the 2 in. mark.

Construction of Triangles.

(*a*) Given the lengths of the three sides.

Suppose we wish to construct a triangle whose sides are 1 in., 2·7 in., 3 in.

(Note. A triangle can be constructed only when any two sides are together greater than the third side.)

Draw the base $AB = 3$ in. (fig. 4). With centre A and radius 1 in. draw an arc with compasses. With centre B and radius 2·7 in. draw a second

arc to cut the first arc at *C*. Join *AC* and *BC*. *ABC* is the required triangle.

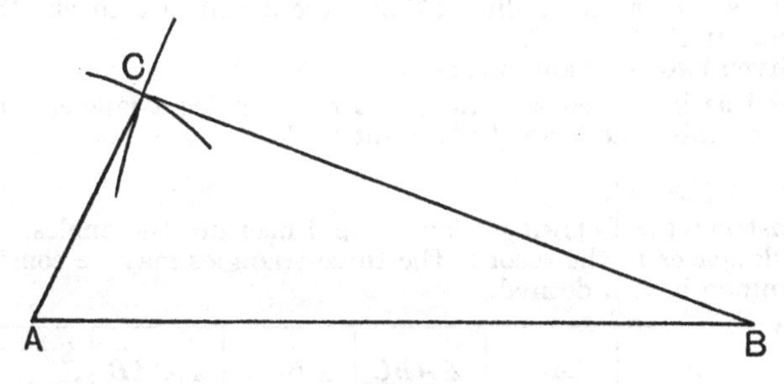

Fig. 4

(*b*) To construct a right-angled triangle given any two sides.

Suppose we are given base 2·5 in. and perpendicular 1·1 in.

Draw base *AB* to required length, 2·5 in. At *B*, with either set square or protractor, erect a perpendicular and mark off *BC* = 1·1 in. Join *AC* (fig. 5).

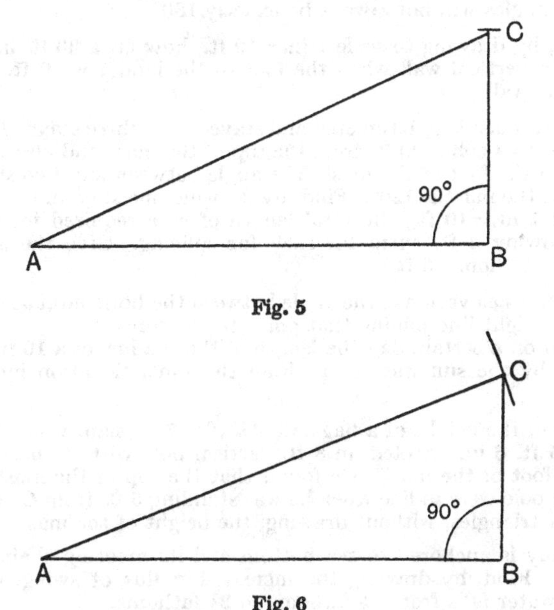

Fig. 5

Fig. 6

If instead of the perpendicular, we are given hypotenuse = 2·7 in., then draw the base as before (fig. 6) and erect the perpendicular. From A as centre and radius 2·7 in. describe an arc to cut the perpendicular at C. Join AC.

(*c*) Given two sides and one angle.

Proceed as in the construction of a right-angled triangle, using the angle given this time instead of a right angle.

EXERCISE XXXVII.

1. Construct the following triangles and measure the angles. Fill up the blank spaces in the record. The three triangles may be constructed on a common base if desired.

AB	BC	AC	$\angle ABC$	$\angle BCA$	$\angle CAB$	Sum of the three angles
3·5	2·6	2·1				
3·5	3·78	0·77				
3·5	1·22	3·36				

NOTE. When working with a protractor to ½° accuracy, the sum of the three angles will not always be exactly 180°.

2. Find, by drawing to scale 1 in. = 10 ft., how far a 30 ft. ladder will reach up a vertical wall when the foot of the ladder is 10 ft. from the base of the wall.

3. A 60 ft. mast is to be erected and stayed with three stays. Each stay is fastened to a collar 20 ft. from the top of the mast and also to a stake 30 ft. from the foot of the mast. The angle between any two stakes and the foot of the mast is 120°. Find, by drawing one stay and the mast to a scale of 1 in. = 10 ft., the total length of wire required for the three stays, allowing 2 ft. extra for each for splicing. Give the answer in fathoms (1 fathom = 6 ft.).

4. The sun's elevation is the angle between the horizontal at any point and the straight line joining that point to the sun.

At noon on a certain day the length of the shadow of a 10 ft. vertical pole cast by the sun was 15 ft. Find the sun's elevation by drawing to scale.

5. To find the height of a flagstaff AB (fig. 7) a man, whose height of eye was 5 ft. 6 in., erected an 8 ft. vertical pole with its base C 20 ft. from the foot of the mast. He found that the top of the mast and the top of the pole were in line when he was standing 5 ft. from C. Calculate by similar triangles, without drawing, the height of the mast.

6. A buoy is anchored to sea bottom and its mooring chain is 25 ft. in length. Find, by drawing the increased radius of swing, when the depth of water falls from 3½ fathoms to 2½ fathoms.

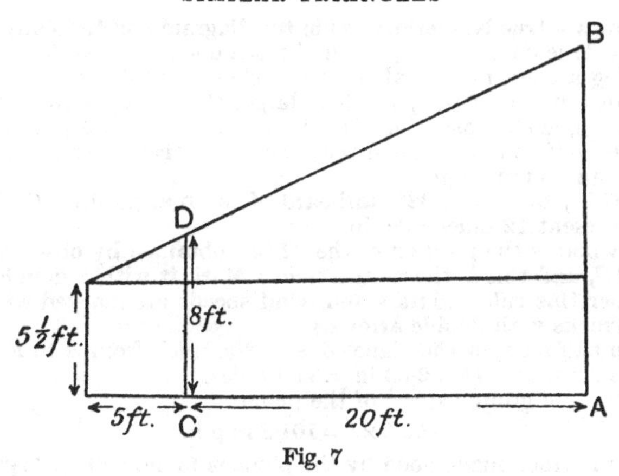

Fig. 7

7. Construct a triangle ABC such that base $AB = 3$ in., $\angle CAB = 62°$, $\angle CBA = 53°$. With the diagonal scale measure and state the lengths of AC and BC.

Simple problems in the construction of triangles occur when finding track and ground speed from a known air speed (A/S), true course and drift.

EXAMPLE. Suppose a plane is flying at A/S 80 m.p.h. on a true course of 067° and at the end of half an hour the drift is found by observation to be 12 miles, with the plane 32° to starboard of her dead reckoning (D.R.) position. Find the track and ground speed (see fig. 8).

Remember that the D.R. position is calculated from A/S and time only on the true course set, and makes no allowance for drift.

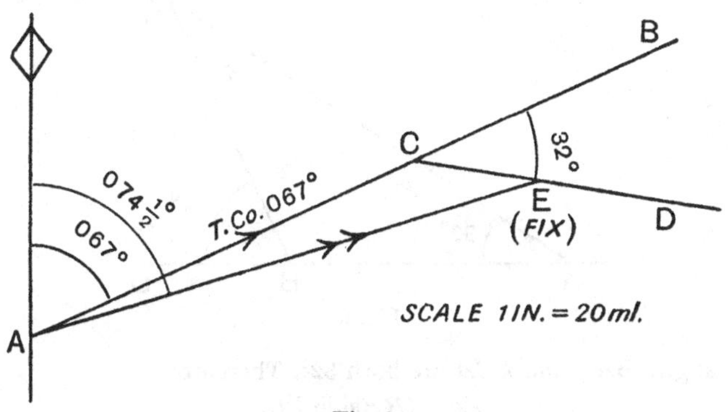

Fig. 8

First draw a true N. meridian as in the diagram and from any point *A* lay off the true course (067°) *AB* and place one arrow on it.

Choosing a convenient scale (the one chosen for fig. 8 is a small one because of the size of the page. The larger the scale the more accurate the working), in this case 1 in. = 20 miles, mark off *AC* = 2 in. to represent 40 miles or half an hour's run at the given A/S. *C* is then the D.R. position after half an hour's flight.

From *C* lay off *CD* at 32° starboard of the D.R. position *C*. Measure *CE* to represent 12 miles = 0·6 in.

Then *E* marks the position of the "Fix" obtained by observation.

Join *AE*, and this is the actual track. Mark it with a double arrow. (Remember this rule: Courses and wind speeds are marked with single arrows. Tracks with double arrows.)

Thus in half an hour the plane covered the track from *A* to *E*.

By measurement *AE* = 2·53 in. = 50·6 miles.

Therefore the ground speed of the plane

$$= 50{\cdot}6 \times 2 = 101{\cdot}2 \text{ m.p.h.}$$

and the true track made good by the plane is found to be 074½° T.

8. Find the ground speed and the track from the following: Air speed 84 m.p.h.; True course 123°; Observation fix after 20 min. flight showing plane to be 10 miles and 25° to port of D.R. position.

9. Find the drift per hour in miles if Air speed is 100 m.p.h.; True course 265°; Drift sight angle 13° port; Ground speed 84 m.p.h.

One property of the triangle that proves of value in coastal navigation is that if any two angles of a triangle are equal, the sides opposite to them are also equal.

Also the exterior angle of any triangle (obtained by producing one side) is equal to the sum of the interior and opposite angles (see fig. 9).

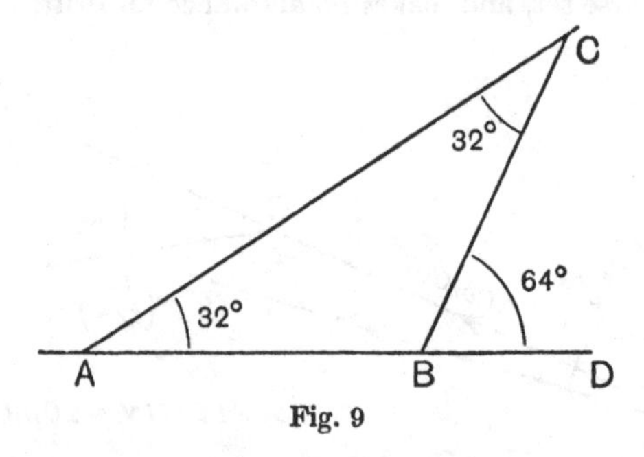

Fig. 9

The angles *BAC* and *BCA* are both 32°. Therefore

side *AB* = side *BC*.

Also when AB is produced to D the exterior angle CBD is equal to the sum of the two interior and opposite angles BAC and BCA.

Stated briefly: Since

$$\angle BAC = \angle BCA,$$

therefore $$AB = BC,$$

and also $$\angle CBD = \angle BAC + \angle BCA = 2\angle BAC.$$

This is made use of to find a vessel's distance from a prominent object, by what is known as "doubling the angle on the bow" (see fig. 10).

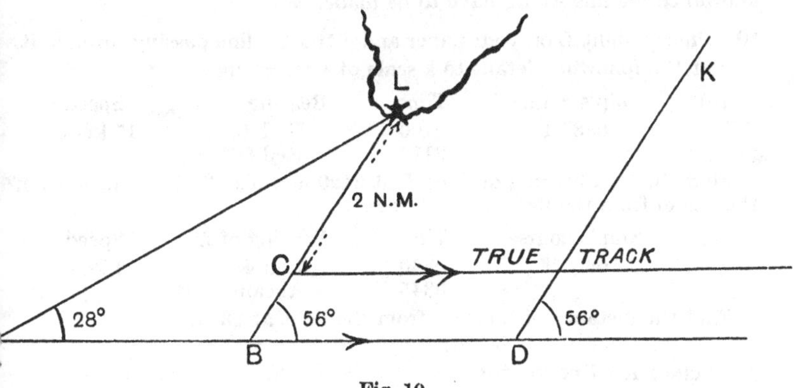

Fig. 10

A vessel is on a course AD steaming at 12 knots. At time 2300 a light L is observed to bear Red 28°.

The bearing instrument is then at once set to twice this angle, i.e. Red 56°, and at 2310 the same light was observed to bear Red 56°. Since the time interval between bearings was 10 min. and the speed 12 knots, the ship must have travelled 2 nautical miles.

We know that $$\angle LBD = \angle LAB + \angle BLA,$$

i.e. $$56° = 28° + \angle BLA.$$

$$\therefore \angle BLA = 28°.$$

Therefore $$BL = BA.$$

But BA is the distance travelled by the vessel in 10 min. = 2 nautical miles.

Therefore, when at B, the ship must be 2 nautical miles from light L in direction LB.

To plot this observation a line DK, making an angle Red 56°, is drawn from any convenient point on the course. With parallel rulers, or set squares, a line LB is drawn parallel to DK. With dividers open to represent 2 nautical miles of chart scale, mark this off along LB from point L. This will bring us (say) to point C.

This point C is called a "Fix" by bearing and distance.

The track line is then drawn through C parallel to the old course.

Another special case of this fix is known as the "four point bearing'.

The first bearing taken here is Red or Green 45° (four compass points) and the second bearing Red or Green 90°, i.e. the object is then "abeam", or at right angles to the ship's fore and aft line.

Remember, in both these examples no notice has been taken of any possible drift due to tide. If there were any tidal drift the fore and aft line of the ship would not have been parallel to the charted course and allowance for this would have to be made.

10. Place a light L on your paper and a true N. line passing through it. Plot the following details to a scale of $\frac{1}{2}$ in. $=1$ mile:

(*a*)	Ship's course	Time	Bearing of L	Speed
	088° T	2100	Red 40°	15 knots
		2112	Red 80°	

Find the Red bearing of light L at 2120 and also find the distance of the vessel from the light at that time.

(*b*)	Ship's course	Time	Bearing of L	Speed
	065° T	0330	Red 45°	18 knots
		0345	Abeam (port)	

Find the distance of the ship from the light at 0355.

Exercises for Protractor.

Exercise XXXVIII.

1. Place a point A at any convenient place on your paper and draw a true N. meridian passing through it.

Using scale 10 miles $=1$ in., lay off from this point the following courses and distances:

	Course	Distance		Course	Distance
AB	073°	35 miles	AF	231°	31 miles
AC	$111\frac{1}{2}$°	$28\frac{1}{2}$ miles	AG	270°	$26\frac{1}{2}$ miles
AD	$177\frac{1}{2}$°	$24\frac{1}{2}$ miles	AH	$301\frac{1}{2}$°	$15\frac{1}{2}$ miles
AE	191°	22 miles	AK	359°	27 miles

Join B, C, D, E, F, G, H and K and record the distances BC, CD, DE, ... KB in miles.

2. Make an exact copy of fig. 11 either by tracing or with compasses. A, B and C represent three charted objects to the scale given and a true meridian is shown. From a ship the following true bearings were taken of these objects consecutively: A bore 338°, B bore 020° and C bore $070\frac{1}{2}$°.

If the reciprocals of these bearings are laid off from their respective objects the position lines will form a small triangle called a "position triangle" (or more familiarly a "cocked hat"). Place a dot in the centre of this triangle for the observed position of the ship and measure the distance of this fix from *A* in miles. This is known as a "three point fix".

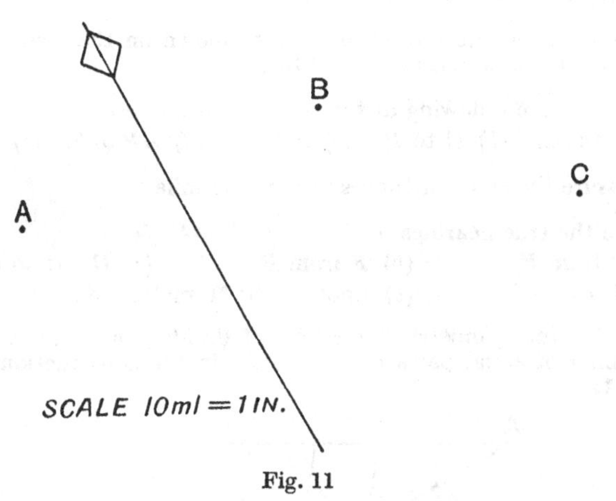

Fig. 11

NOTE. Had the ship been stationary and the bearings taken with absolute accuracy the fix would have been represented by a point and not a small triangle.

SCALES

Turn to the Chart No. 3 on p. 64.

On the left-hand border of this chart two scales of distance are shown:

(i) Scale of statute miles at 5 mile intervals.

(ii) Latitude scale in minutes or nautical miles.

If you measure carefully with your dividers you will find that the length of the ten lat. minutes from lat. 52° 50′ N. to lat. 53° N. is greater than the ten lat. minutes from lat. 51° N. to lat. 51° 10′ N. The same applies to the scale of statute miles.

This is always the case on a Mercator Chart. So, therefore, we must take care to measure distances between objects from that portion of the scale that concerns them.

The distance from *OA* to *NT* is found by joining *OA* to *NT* and then opening the dividers to 10 nautical miles (or statute miles if we want them) as given on the scale between lat. 52° 20′ N. and lat. 52° 30′ N.

Check that this distance in N.M. and S.M. is approximately 50 and 61 respectively.

EXERCISE XXXIX.

1. From the Chart No. 3 give the latitude and longitude of (*a*) *K*, (*b*) *BO*, (*c*) *U*, (*d*) *OA*, (*e*) *B*.

2. From the two dotted isogonal lines shown on the chart find what variation you would use at **G* in 1941.

3. Measure the following distances in statute miles:
(*a*) *E* to *M*. (*b*) *H* to *P*. (*c*) *F* to *R*. (*d*) *AP* to *S*. (*e*) *A* to *OA*.

4. Measure the same distances in nautical miles.

5. Give the true bearings of:
(*a*) *B* from *F*. (*b*) *F* from *B*. (*c*) *OA* from *Q*.
(*d*) *S* from *AP*. (*e*) Spot height 70 metres from *O*.

The following geometrical method of dividing a straight line into any number of equal parts is very useful in the construction of scales (see fig. 12).

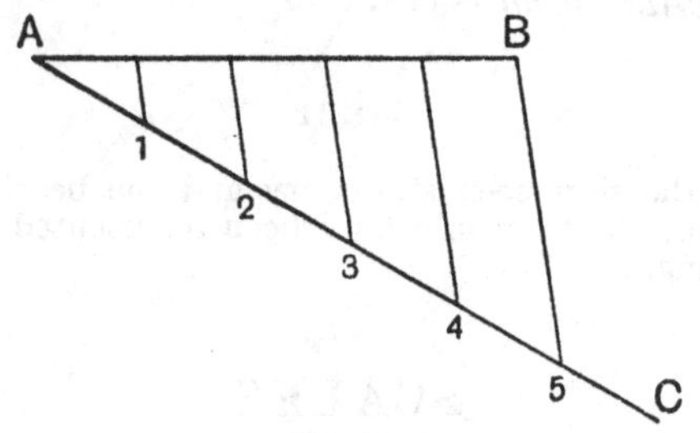

Fig. 12

Let *AB* be a line to be divided into 5 equal parts.

From *A* draw any line *AC* at any convenient angle to *AB*. With compasses mark off 5 equal divisions of convenient length along *AC*, and number them.

Join no. 5 to *B* and through 4, 3, 2 and 1 draw parallels to *B*5.

AB is then divided into 5 equal parts.

Scales for use with particular maps.

Consider a map of R.F. $\frac{1}{36000}$, i.e. 2 in. = 1 nautical mile of 2000 yd.

Distances on such a map may be measured by an inch ruler and the corresponding ground distance may be obtained by calculation. This is laborious and likely to lead to error.

It is far more convenient to construct a special scale for use with this map, so that ground distances can be read directly from it. This is how it is done.

EXAMPLE 1. To construct a scale of 2 in. = 1 nautical mile (of 2000 yd.) to show accuracy of 200 yd. or 1 cable.

(NOTE. 1 cable = 0·1 N.M. = 200 yd. = 100 fathoms.)

Draw a line 4 in. long to represent 2 N.M. (fig. 13). Divide this into 2 equal parts each 2 in. long and each representing 1 N.M.

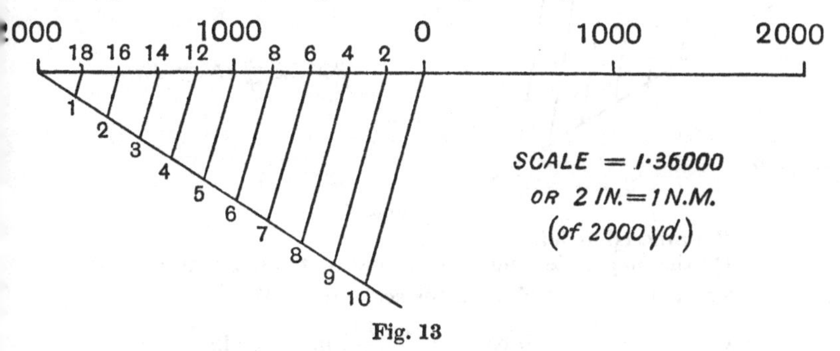

Fig. 13

Mark the zero in the middle and subdivide the left hand 2 in. The accuracy required is 200 yd. (1 cable or $\frac{1}{10}$ of 1 N.M.).

So, by parallels, the left hand 2 in. is divided into 10 equal parts, each of which represents 200 yd. on the ground.

EXAMPLE 2. Fig. 14 shows a scale suitable for use with any map whose R.F. is 1 : 1,000,000, i.e. 15·78 miles = 1 in.

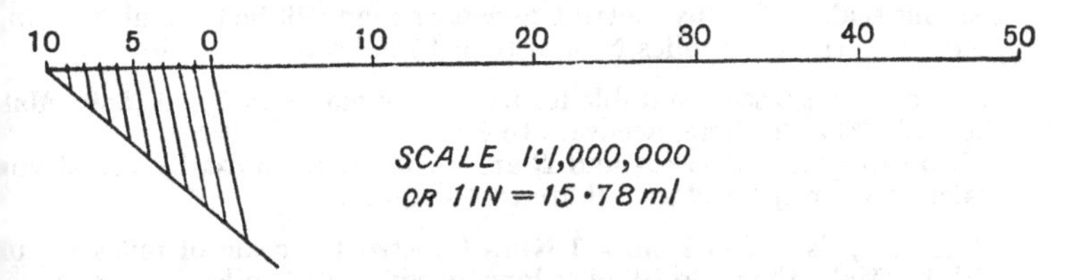

Fig. 14

In this scale we are dividing into 10 mile lengths with one 10 mile group to the left of zero subdivided into single miles.

The 10 mile length on this scale will be

$$\left(\frac{10}{15{\cdot}78}\right) \text{ in.} = 0{\cdot}63 \text{ in.}$$

and this is accurately measured off from the diagonal scale of inches before we start (fig. 3).

Example 3. Suppose we desire a scale of Kilometres to use with map whose English scale is $\frac{1}{4}$ in. = 1 mile with accuracy to 1 Km. (fig. 15

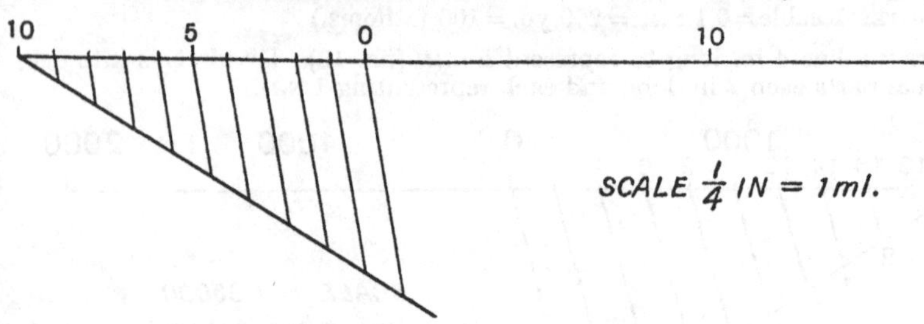

Fig. 15

We proceed as follows:

On the map scale 1 mile = $\frac{1}{4}$ in. and we know 1 Km. = $\frac{41}{66}$ miles.

Therefore 1 Km. on the scale = $(\frac{41}{66} \times \frac{1}{4})$ in.

and 10 Km. = $\left(\frac{41 \times 10}{66 \times 4}\right)$ in. = 1·55 in.

Using the diagonal scale again we mark off 10 Km. lengths eac 1·55 in. long and subdivide the left-hand one, from zero, into 10 equ parts, each representing 1 Km.

Exercise XL.

1. Construct a scale of miles for use with an Ordnance Survey map scale 1 in. = 1 mile. Make your scale 5 in. long accurate to 1 furlon Use this scale to find by plotting how far a man will be from his startin point if he travels 2 miles N. and then $1\frac{1}{2}$ miles E.

2. Construct a scale suitable for use with a map R.F. 1 : 506,880. Mal the scale 30 miles long, accurate to $\frac{1}{2}$ mile.

If on this map places *A* and *B* are 2·7 in. apart state, by use of yo scale, how far apart *A* and *B* are on the ground.

3. A map is scaled 1 cm. = 1 Km. Construct a scale of miles for u with it. Make the scale 10 miles long, accurate to 1 mile. Use your sca to convert 10 Km. into miles.

4. Two places on a map are 3·1 in. apart. On the ground they a 4·5 miles apart. Construct a scale of miles for use with this map, showi a full length of 7 miles and accuracy to 0·1 mile. Find the ground distan between *A* and *B* if the map distance between them is 3 in.

5. Construct a scale for use with a map of R.F. 1 : 2,500,000. Ma your scale 160 miles long with accuracy to 2·5 miles. Use your scale find the ground distance between two places that are 1·8 in. apart c the map.

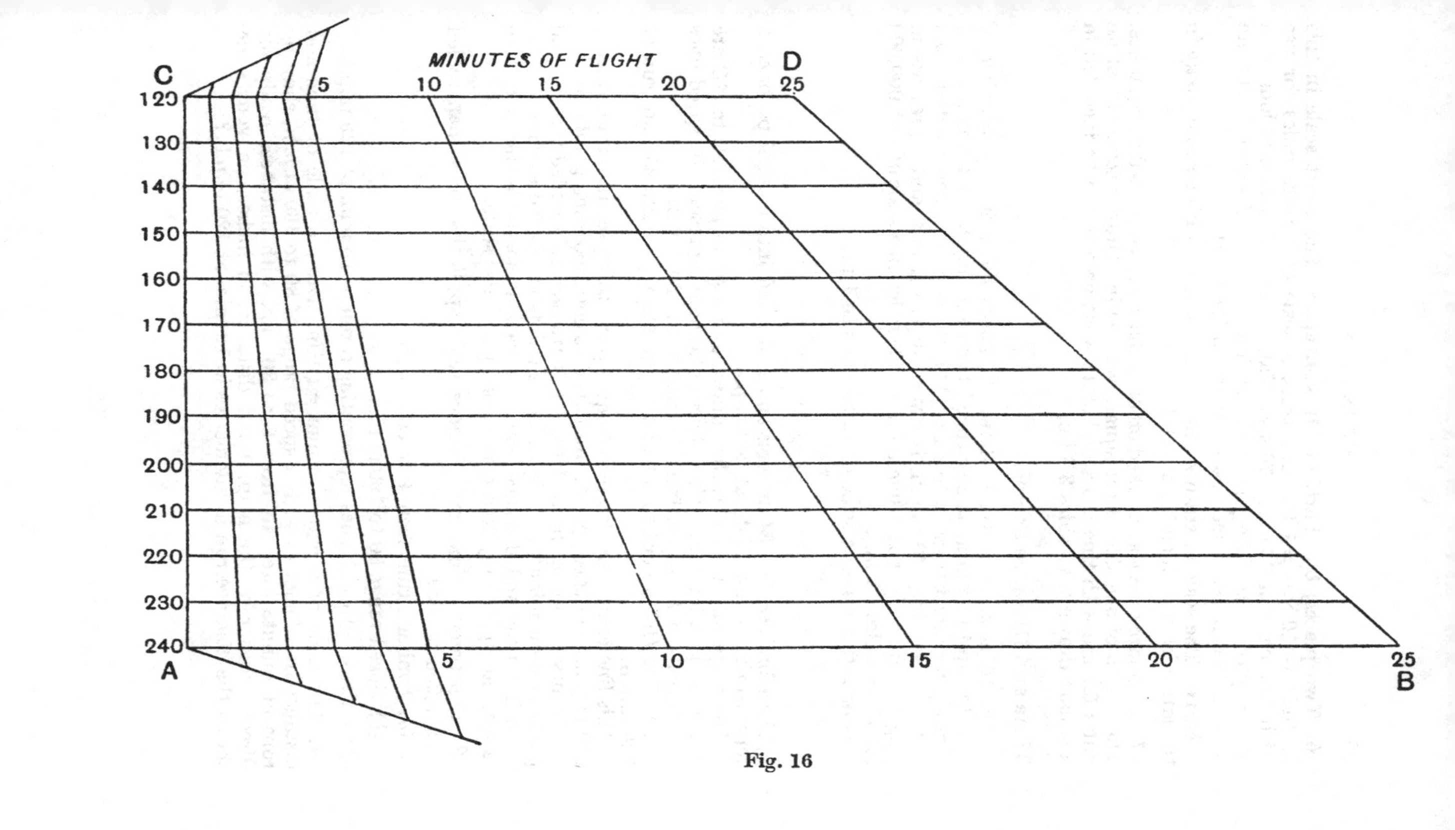
MINUTES OF FLIGHT
C
D
A
B
120
130
140
150
160
170
180
190
200
210
220
230
240
5
10
15
20
25
5
10
15
20
25

Fig. 16

6. Two places, by chart, are 18 N.M. apart. The chart scale in this latitude is 10 N.M. = 1 in. Construct a scale of statute miles for use with this chart if 66 N.M. = 76 S.M. Make your scale 50 miles long with accuracy of 1 S.M. Use your scale to determine the distance in statute miles between two places 2·4 in. apart on the chart.

NOTE. The scale of statute miles could be used with accuracy *only* in the latitudes for which 1 N.M. = 0·1 in.

7. A square area was marked off on a chart and contained 64 sq. miles. The length of one side of the square was 2·4 in. Draw a scale of statute miles for use with this chart and state what area would be included in a square drawn with sides 3·6 in.

Time and Distance Scales.

Fig. 16 is a scale which shows distances flown in any number of minutes up to 25 min. at any speed from 120 m.p.h. to 240 m.p.h. and is ready for direct transfer to a 1 : 1,000,000 map. It is constructed thus:

The speed line 240 m.p.h. is drawn first, i.e. *AB*, and represents in length 25 min. flight at 240 m.p.h., i.e. 100 miles on a scale of 1 : 1,000,000 (or 15·78 miles = 1 in.).

Now 100 miles on a scale of 15·78 miles to 1 in. is

$$\frac{100}{15{\cdot}78} \text{ in.} = 6{\cdot}34 \text{ in.}$$

Therefore *AB* is 6·34 in. long and is divided into 5 equal parts each representing 5 min. flight at 240 m.p.h.

The perpendicular *AC* is then drawn and a set of parallels to *AB* are drawn at $\frac{1}{4}$ in. intervals. These horizontal lines represent the speed lines at intervals of 10 m.p.h. from 240 m.p.h. to 120 m.p.h.

The line *CD* is marked off to 3·17 in. and represents 25 min. run at 120 m.p.h.

This line is also divided into 5 equal parts to represent 5 min. runs at 120 m.p.h., and these points are joined to corresponding points on *AB*.

The first 5 min. of both *AB* and *CD* are each subdivided into 5 equal parts to show minutes of flight. These points are joined as shown.

Such a time and distance scale, for various speeds within a certain range, can apply only to a map of one particular scale.

8. Construct a time and distance scale applicable to a map scaled 1 : 500,000 to show

(*a*) 1 minute runs from 0 to 30 min.

(*b*) Speed range 80 to 160 m.p.h.

State the length in inches representing a run of 16 min. at 120 m.p.h.

9. The scale of a map is 1 mm. = 1 Km. Draw a scale of time and distance in kilometres with a speed range of 80 to 160 *m.p.h.*, showing runs at minute intervals from 0 to 30 min. with distances ready for direct transfer to the map. What distance, in inches, on your scale shows the distance run in kilometres in 12 min. at 120 m.p.h.?

10. A map has a scale, in inches, 1 : 250,000. Draw a time and distance scale with a speed range of 90 to 180 Km. per hour, giving distances in miles at any minute from 0 to 30 min. ready for direct transfer to the map. What distance, in inches, on your scale shows the mileage for a 12 min. run at 120 Km. per hour?

PARALLELOGRAM AND TRIANGLE OF VELOCITIES

Scalar and Vector Quantities.

If a ship is steaming at 12 knots and we represent this speed by a scale of 1 knot = 0·1 in., then any line 1·2 in. in length will represent the speed of the ship.

Such a line is called a *Scalar* quantity because it is drawn to scale.

This line does not take into account one important feature of the ship's speed. It does not indicate the direction of the ship's speed.

Suppose the ship in question was steaming N.E. at 12 knots.

Now if we draw a true N. meridian and from some point on it we draw a line 1·2 in. long in a direction N.E. from the meridian, then this line will represent both the *speed* of the ship and its *direction*.

Such a line is called a *Vector* quantity and it represents the true velocity of the ship.

Parallelogram of Velocities.

Any four-sided figure with its opposite sides equal and parallel is called a *Parallelogram.*

A diagonal of a parallelogram divides the figure into two identically equal triangles.

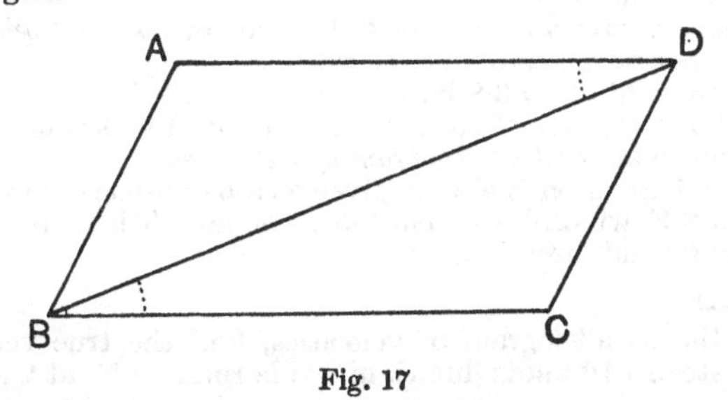

Fig. 17

Fig. 17 is a parallelogram. *AB* is equal and parallel to *CD*. *BC* is equal and parallel to *AD*. *BD* is a diagonal and △ *ABD* is identically equal to △ *CDB*.

The angles *ADB* and *CBD* are called "alternate" angles, and are equal to one another.

Example 1. Suppose now, a ship, steaming due E. at 12 knots, is sailing in an ocean current, i.e. a body of water in motion, which is flowing N.E. at 4 knots.

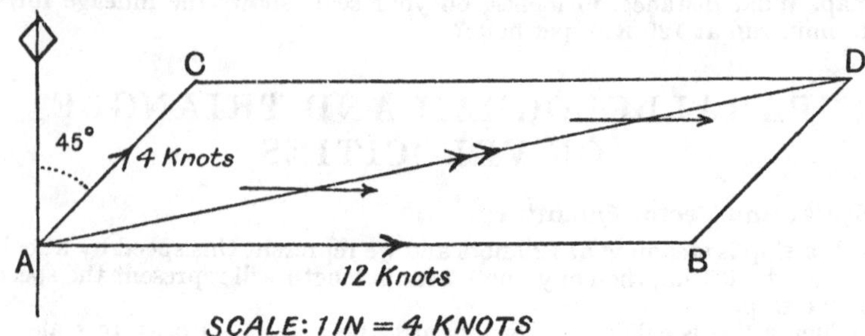

Fig. 18

In fig. 18, using a scale 4 knots = 1 in., *AB* is the vector representing the ship's velocity (speed and direction) and *AC* is the vector representing the velocity of the tide.

In still water the ship would have travelled from *A* to *B* in one hour, but, since the tide will also carry the ship with it, we shall find that at the end of one hour the ship has arrived at *D*, where the figure *ABCD* is a parallelogram.

Actually, also, we shall find that the vessel reached *D* along the track *AD*, and was never off the track although the ship's fore and aft line was always parallel to *AB*.

Also the distance travelled by the ship was from *A* to *D* in one hour. This distance is greater than 12 nautical miles because, in this case, the tide has helped the ship along.

AD is also a vector and represents the true or *resultant velocity* of the ship, due to her own steam and the tide.

By measurement $AD = 3{\cdot}8$ in. and bearing 082° T.

The resultant velocity of the ship is therefore 15·2 knots 082° T.

This is known as the *Parallelogram of Velocities.*

Note. Wind direction is always given as a *true* direction *from* which it blows, e.g. a S.E. wind blows from 135°. The label T is therefore omitted when defining wind directions.

Exercise XLI.

1. Using the parallelogram of velocities, find the true velocity of a ship which steams 10 knots due E. in a tide running N. at 2 knots.

2. A ship is steaming S.W. at 12 knots and the tide is running N. at 3 knots. Determine the true velocity of the ship.

3. A plane is flying on a course 078° T at 90 m.p.h. and the wind is blowing *from* 357° at 27 m.p.h. Find the ground speed and ground track of the plane. Use a scale of 18 miles = 1 in.

4. Work the following from details given. Find ground speed and track.

	True course of plane	Speed	Wind direction	Wind Speed
(a)	196°	120 m.p.h.	301°	40 m.p.h.
(b)	275°	100 m.p.h.	182°	38 m.p.h.
(c)	271°	90 m.p.h.	265°	37 m.p.h.
(d)	195°	80 m.p.h.	090°	36 m.p.h.
(e)	002°	85 m.p.h.	010°	25 m.p.h.
(f)	075°	96 knots	250°	32 knots
(g)	123°	93 knots	050°	28 knots

Triangle of Velocities.

EXAMPLE. To find the velocity of the wind, when true course, air speed, track and ground speed are known.

Let us examine this problem from a specific case. A plane whose true course is 067° and air speed 78 m.p.h. makes a track of 086° T and a ground speed of 90 m.p.h. (fig. 19).

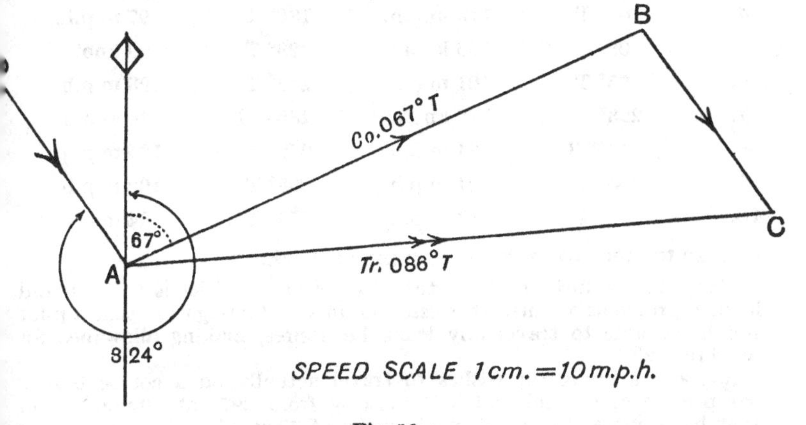

Fig. 19

First draw a true meridian and from any convenient point *A* on the meridian lay off *AB* 067°. Choose a speed scale, say 1 cm. = 10 m.p.h., and write the scale on the drawing. Then mark off the length *AB* = 7·8 cm. (to represent 78 m.p.h.) and "arrow" it. This line *AB* is the vector of air speed and true course.

From *A* also lay off *AC* (086°) and *AC* = 9 cm. This line *AC* is the vector of the ground speed and track. Mark this with a "double arrow".

By dead reckoning the plane according to air speed and true course steered should be at B at the end of one hour. By observation the plane is at C at the end of one hour. The conclusion from this is that during one hour the wind carried the plane a distance $= BC$ and in the direction from B to C.

BC, therefore, is a vector of the wind velocity. To state the direction of the wind a line is drawn parallel to BC and passing through point A. From A a length AD is marked off $= BC$.

Measuring the angle through E., S. and W., we obtain the wind direction as 324°. And from length of AD (3 cm.) and the scale, 1 cm. = 10 m.p.h., the speed = 30 m.p.h.

In this example we have made use of the *Triangle of Velocities*.

EXERCISE XLII.

Find the wind speed and direction from these tabulated details:

	True course	Air speed	Track	Ground speed
1.	200° T	103 m.p.h.	184° T	83 m.p.h.
2.	069° T	120 m.p.h.	057° T	100 m.p.h.
3.	121½° T	95 m.p.h.	098° T	71 m.p.h.
4.	161° T	113 m.p.h.	180° T	97 m.p.h.
5.	308° T	66 knots	328° T	99 knots
6.	286° T	101 m.p.h.	272° T	93 m.p.h.
7.	223° T	115 knots	236½° T	78 knots
8.	048½° T	94 m.p.h.	052° T	125 m.p.h.
9.	090° T	101 m.p.h.	068° T	102 m.p.h.
10.	186° T	88 m.p.h.	155° T	94 m.p.h.

Course to steer to make good a given track.

EXAMPLE. A further use of the triangle of velocities is to be found in such problems as this. It is essential in aerial navigation that a pilot should be able to travel any track he desires, making allowance for wind (fig. 20).

Suppose a navigator wishes to travel actually on a course 072° T irrespective of a wind which is blowing from 297° at 40 m.p.h. and that he proposes to fly at an air speed of 73 m.p.h. The procedure is as follows.

Draw a true meridian TA through the starting position A.

Lay off AE, any length at 072° T, to represent the track to be made good. Put two arrows on it.

Now from A draw AB to represent one hour's movement of the wind to a scale of (say) 1 cm. = 10 m.p.h. (Note, the direction of A to B is the reciprocal of 297°.) AB is therefore the vector of the wind velocity.

With compasses open to 7·3 cm. (73 m.p.h. on the scale) and with B as centre, strike an arc cutting the track at C. Then BC is the true course to steer, and as in our figure it starts from *B* instead of *A*, we then transfer it to *A* by parallels, making $AD = BC$.

Thus *AD* (by measurement 048° T) is the vector of the air speed and course to steer, which will, owing to the wind velocity, carry the plane along the track *AE* required.

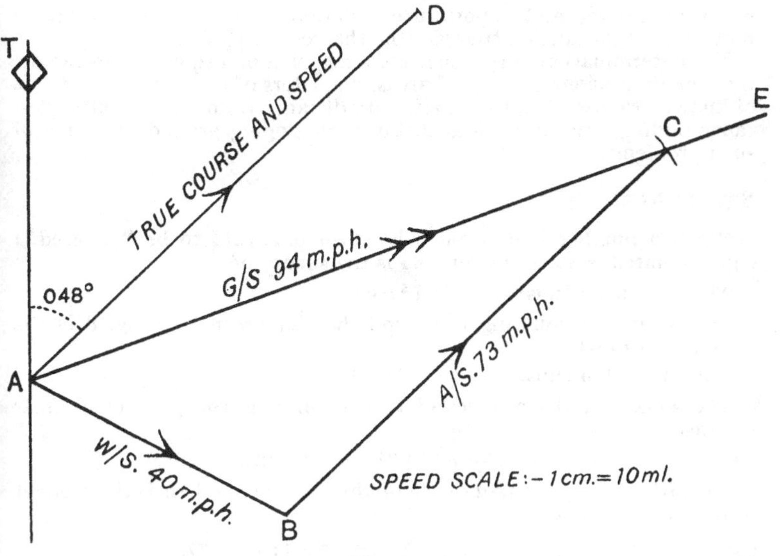

Fig. 20

EXERCISE XLIII.

Find the course to steer and the ground speed of the plane in the following cases:

	Track to make good	Wind		Air speed
1.	296° T	053°	34 m.p.h.	100 m.p.h.
2.	090° T	224°	36 m.p.h.	93 m.p.h.
3.	070° T	322°	47 m.p.h.	121 m.p.h.
4.	258½° T	139°	42 m.p.h.	99 m.p.h.
5.	147° T	029°	32 knots	104 knots
6.	242½° T	304°	40 m.p.h.	140 m.p.h.
7.	286° T	346°	37 m.p.h.	135 m.p.h.
8.	350° T	267°	31 knots	115 knots
9.	091° T	016°	30 m.p.h.	128 m.p.h.
10.	096° T	053°	32 knots	127 knots

SQUARE ROOT AND RIGHT-ANGLED TRIANGLE

Many interesting and important calculations involve the use of square root or the right-angled triangle (the theorem of Pythagoras).

The determination of the length of a diagonal of a square or rectangle, the true dimensions of mapped areas, diameters of circular wires or rods of known cross-sectional area, visibility distance from different altitudes, range-finding, true horizontal distance on sloping ground are some of such problems.

Square Root.

When a number is multiplied by itself it is said to be "squared": e.g. 8 squared is $8 \times 8 = 64$, and 64 is the square of 8.

This is written thus: $8^2 = 64$.

The reverse of squaring is to find the "square root": e.g. 8 is the square root of 64.

This is written thus: $\sqrt{64} = 8$.

The square of the product of two numbers is the product of their squares:

e.g. $(5 \times 3)^2 = 5^2 \times 3^2 = 25 \times 9 = 225$.

Similarly, the square root of the product of two numbers is the product of their square roots:

e.g. $\sqrt{121 \times 49} = \sqrt{11 \times 11 \times 7 \times 7} = 11 \times 7 = 77$.

Similarly with fractions, thus

$$\sqrt{\frac{25}{64}} = \sqrt{\frac{5 \times 5}{8 \times 8}} = \frac{5}{8}.$$

By ordinary multiplication we can draw up a table of the squares of the numbers from 1 to 20:

Number	Square	Number	Square
1	1	11	121
2	4	12	144
3	9	13	169
4	16	14	196
5	25	15	225
6	36	16	256
7	49	17	289
8	64	18	324
9	81	19	361
10	100	20	400

It is at once clear from this table that most numbers have no *exact* square root, but we can calculate an *approximate* answer.

For example, what is the square root of 2?

We know that $\sqrt{1}=1$ and $\sqrt{4}=2$ so that $\sqrt{2}$ must lie between 1 and 2.

From the above table we see that $14^2=196$ so that $\sqrt{196}=14$.

And
$$\sqrt{\frac{196}{100}}=\frac{14}{10}=1{\cdot}4.$$

But $\sqrt{\frac{196}{100}}=\sqrt{1{\cdot}96}$, which is very nearly $\sqrt{2}$.

Therefore $\sqrt{2}$ is a little greater than 1·4.

A still closer approximation is 1·414.

Here, for reference, is a table of square roots, correct to three places of decimals, of the numbers from 1 to 10:

Number	Square root	Number	Square root
1	1	6	2·449
2	1·414	7	2·646
3	1·732	8	2·828
4	2	9	3
5	2·236	10	3·162

The easiest and most convenient way to find the square root of any number is by the use of logarithms (see p. 120).

Square root is used to determine the length of the side of a square from its area.

For example, a square of area 16 sq. in. has a side $\sqrt{16}=4$ in. long.

In the same way if it is known that on a map a square whose side is of certain length represents a certain area on the ground, then the scale of the map can be determined.

Example. On a certain map a square of 2 in. side marks off an area representing 16 sq. miles on the ground. What is the scale of the map in miles per inch?

The scale is:

Length of side of "ground" square (in miles) to length of side of "map" square (in inches), i.e. $\sqrt{16}$ miles to 2 in.

Which is 4 miles to 2 in.

The scale of the map, therefore, is 2 miles to 1 in.

EXERCISE XLIV.

1. On a certain map a square with 3 in. sides marks off an area representing 81 sq. miles on the ground. What is the scale of the map in miles per inch?

2. What is the scale of a map on which a square with 2 in. side represents an area on the ground of 144 sq. miles?

3. What is the scale of a map on which a square with 3 ft. side represents an area of 324 sq. miles on the ground?

4. What is the scale of a map on which a square with side $\frac{3}{4}$ in. in length represents an area of $5\frac{1}{16}$ sq. miles on the ground?

5. By using the method shown in the example for finding the approximate value of $\sqrt{2}$, show that (*a*) $\sqrt{13}$ is nearly equal to 3·6, (*b*) $\sqrt{22}$ is nearly equal to 4·7.

The Right-angled Triangle.

In any right-angled triangle the side opposite the right angle is called the "Hypotenuse". The other two sides are the "Base" and the "Perpendicular" (see fig. *a*).

If squares are drawn on the three sides of a right-angled triangle, then

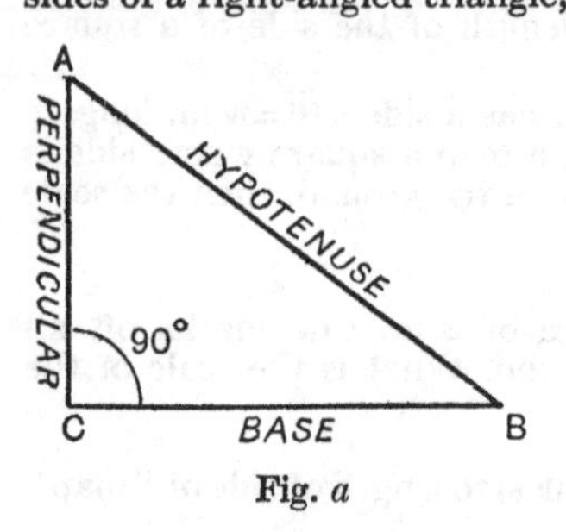

Fig. *a*

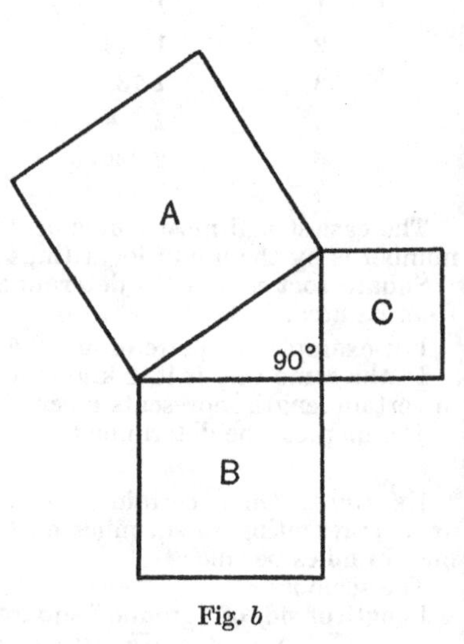

Fig. *b*

the square on the hypotenuse is exactly equal in area to the sum of the other two squares, i.e.

Square A = Square B + Square C (fig. *b*).

This is the theorem of Pythagoras.

This, from fig. 1, may be expressed:

$$AB^2 = AC^2 + BC^2.$$

By transposing:

$$AC^2 = AB^2 - BC^2$$

and

$$BC^2 = AB^2 - AC^2.$$

Thus, by calculating the necessary square root, the length of any side of a right-angled triangle can be found if the length of each of the other two sides is known.

EXAMPLE. The length of the hypotenuse of a certain right-angled triangle is 13 in. If the base is 5 in., what is the length of the perpendicular?

We know that:

$$\text{Hypotenuse}^2 = \text{Perpendicular}^2 + \text{Base}^2.$$

Therefore

$$\text{Perpendicular}^2 = \text{Hypotenuse}^2 - \text{Base}^2,$$

i.e.

$$\begin{aligned}\text{Perpendicular}^2 &= 13^2 - 5^2\\ &= 169 - 25\\ &= 144 \text{ sq. in.}\end{aligned}$$

Thus the perpendicular $= \sqrt{144} = 12$ in.

A particular case of the right-angled triangle is one in which the base and perpendicular are of equal length.

In other words, the hypotenuse in such cases as this is a diagonal of a square.

EXAMPLE. What is the length of a diagonal of a square in terms of the length of its side?

Suppose the length of a side of the square is a in.

Then, in this case,

$$\text{Base} = \text{Perpendicular} = a \text{ in.}$$

So that,

$$\begin{aligned}\text{Hypotenuse}^2 &= a^2 + a^2\\ &= 2a^2 \text{ sq. in.}\end{aligned}$$

And

$$\begin{aligned}\text{Hypotenuse (or diagonal)} &= \sqrt{2} \times a \text{ in.}\\ &= 1{\cdot}414 \times a \text{ in. (approx.).}\end{aligned}$$

Therefore, the approximate length of a diagonal of any square may be found, by multiplying the length of a side by 1·414.

EXERCISE XLV.

1. What is the approximate length of a diagonal of a square with the following sides?

(a) 3 in. (b) 5 in. (c) 11 cm.

2. What is the approximate length of a diagonal of a square of the following area?

(a) 169 sq. in. (b) 2·89 sq. in. (c) 5 sq. in.
(d) 700 sq. ft. (e) $56\frac{1}{4}$ sq. in.

3. Find the missing side of the following right-angled triangles:

	Hypotenuse	Perpendicular	Base
(a)	5	—	3
(b)	41	40	—
(c)	—	12	9
(d)	39	36	—
(e)	$10\frac{1}{4}$	—	$2\frac{1}{4}$

4. A flight of straight stairs is made up of steps each 1 ft. wide by 9 in. high. What is the distance in a straight line from the bottom of the 1st step to the bottom of the 51st step?

5. How far away from the base of a wall must the bottom of a 26 ft. ladder be placed so that it just reaches a window-sill 24 ft. above ground?

6. From a height of 60 ft. up a flagstaff two wire stays are carried to the ground each 25 ft. from the base of the staff. What total length of wire will be needed, allowing 5 % extra for splicing?

7. Roof trusses, of dimensions shown in the figure, support a roof 30 ft. long. What length of timber 6 in. wide will be required to close-board the roof?

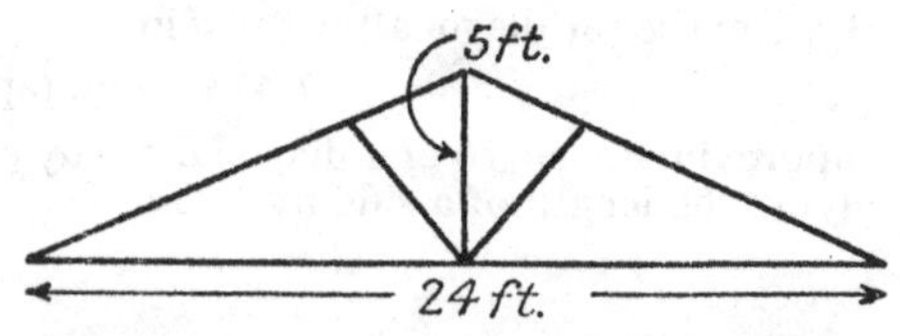

8. A barrage balloon is at an altitude of 2000 ft. The point on the ground vertically below the balloon is 500 yd. from the point where the balloon is anchored. The mooring wire is not taut. Find the length of the mooring wire, allowing 10 % extra for curvature due to slackness.

9. A barrage balloon is flying at an altitude of 4000 ft. The point on the ground immediately below the balloon is 300 yd. from where the balloon is anchored. The mooring cable is not taut and 400 ft. of extra wire are used in this "slackness". If the winding drum can pull in the balloon at the rate of 25 ft. in 2 sec., how long will it take to wind in the balloon?

10. NOTE. When an aircraft is flying in a direction both northwards and eastwards (i.e. on a track between 000° T and 090° T) it is said to be "making northing and easting".

A pilot in an aircraft flying at 246 m.p.h. on a course 012° T finds his northing at the end of half an hour's flight to be 120 miles. What is his easting?

11. An aircraft flying at 260 m.p.h. on a course of 157° T has made good an easting of 25 miles after 15 min. flight. What is the southing?

12. A pilot after flying for 12 min. at 250 m.p.h. on a course of 217° T finds his southing to be 40 miles. He then alters course to 323° T for a further 30 min. at the same speed and finds his new northing to be 100 miles. What is his total westing?

13. A flagstaff is 65 ft. high. From a collar, 20 ft. from the top of the staff, four wire stays are drawn taut to bolts in the ground. Each bolt is 15 yd. from the base of the flagstaff. What length of wire (to nearest foot) is needed for the stays, allowing 2 % extra for splicing? (Assume $\sqrt{2}=1{\cdot}414$.)

14. If $S=1{\cdot}225\sqrt{H}$ is a formula showing the visibility distance in miles at an altitude H feet, what is the visibility distance at 900 feet?

15. What would be the altitude at which visibility is 61·25 miles?

16. An observation post A, on a mountain top, and a battery B, in a valley, are 1·25 in. apart on a map of scale $\frac{1}{63360}$. The report of a gun fired at B takes 6·5 sec. to reach A. What is the height of A if B is 250 ft. above sea level? (Sound travels at 1100 ft. per sec.)

PART TWO

RADIAN (OR CIRCULAR) MEASURE OF ANGLES

A piece of wire, equal in length to the radius of any circle, is bent so as to coincide with a part of the circumference of that circle.

Then the angle which that arc (equal in length to the radius) subtends at the centre of the circle is called a Radian

$$=1^c.$$

We already know that the radius of a circle divides into the circumference 2π times, or into half the circumference π times.

Thus

$180°$ (angle subtended at centre) $=\pi^c$,

$180° = 3{\cdot}14^c$,

$1° = 0{\cdot}01745$ radians,

and $1^c = 57{\cdot}2958°$.

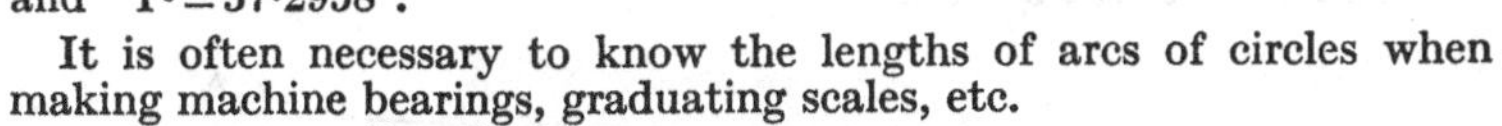

It is often necessary to know the lengths of arcs of circles when making machine bearings, graduating scales, etc.

We know that $\dfrac{\text{Length of arc}}{\text{Radius}} = \text{Angle in radians}$, i.e.

$$\text{Length of arc} = \text{Angle in radians} \times \text{radius}.$$

EXAMPLE. What is the length of an arc of a right angle of a circle of radius $1\frac{1}{2}$ in.?

$$\begin{aligned}\text{Length of arc} &= \text{Angle in radians} \times \text{Radius}\\ &= (90° \times 0{\cdot}01745)^c \times 1{\cdot}5\\ &= 1{\cdot}57 \times 1{\cdot}5 = 2{\cdot}355 \text{ in.}\end{aligned}$$

EXERCISE XLVI.

1. Convert these angles into radians:

(*a*) 100°. (*b*) 28°. (*c*) 5°. (*d*) 3° 30′.
(*e*) 1° 37′. (*f*) 2° 42′. (*g*) 3° 7′ 42″.

2. Convert these angles of circular measure into sexagesimal units:

(*a*) $1{\cdot}1^c$. (*b*) $2{\cdot}534^c$. (*c*) $0{\cdot}00001^c$.
(*d*) $0{\cdot}00005^c$. (*e*) $0{\cdot}02347^c$.

3. The moving needle of an instrument is free to swing through 118° and the needle is $4\frac{3}{4}$ in. long. A blank scale has to be graduated into 20 equal divisions. Find the arc length of each division to the nearest $\frac{1}{1000}$ in.

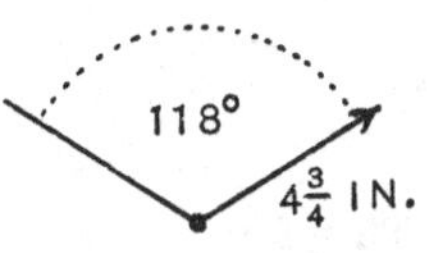

4. A semicircular starboard "bearing-plate" of diameter 18 in. is to be made. Find the length of each of the 180 Green degrees around the arc (to nearest $\frac{1}{1000}$ in.).

5. A drift sight graduated in 360° has each graduation of length 0·105 in. around the outside of the arc. Find the diameter of the brass scale ring.

6. The distance of visibility at an altitude of 2000 ft. is 53 miles. Find the length of the arc of the horizon which subtends an angle of 1° at this range.

Radian Measure of Small Angles.

Circular measure is most useful when dealing with angles of 4° or less. With large angles, like ACB in the figure, it is obvious at once that the straight chord AB is shorter in length than the arc AB.

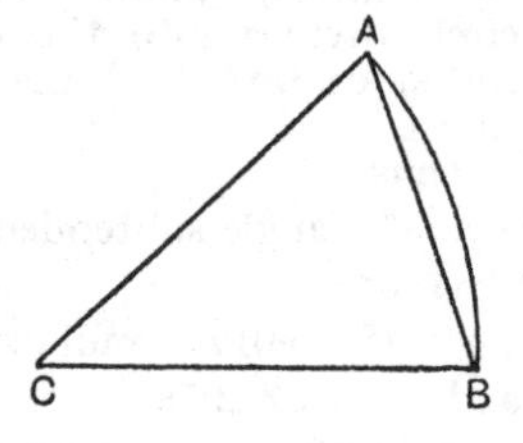

But in the next figure, using a very small angle (4° or less), we can hardly measure the difference between the length of the chord AB and the length of the arc AB.

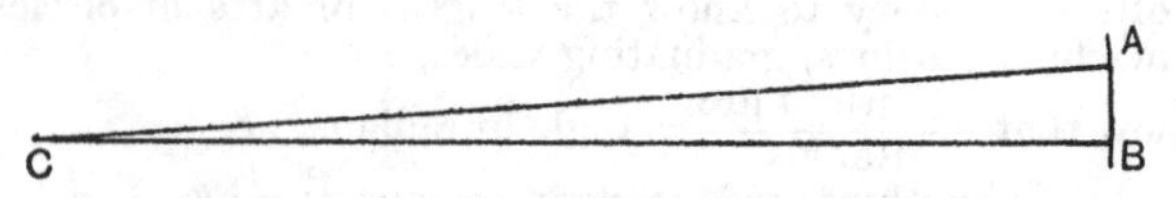

In the diagram the chord and the arc have been carefully drawn, but they overlap so that one cannot be distinguished from the other.

Thus, for angles of 4° or less,

$$\text{Angle in radians} = \frac{\text{Arc}}{\text{Radius}} = \frac{\text{Chord}}{\text{Radius}}.$$

This means that the straight line length AB may be used instead of arc length for angles under 4°.

Example. A ship at sea observes a distant cliff to subtend an angle of 3° by sextant. The chart shows the height of the cliff to be 150 ft. How far off shore is the ship?

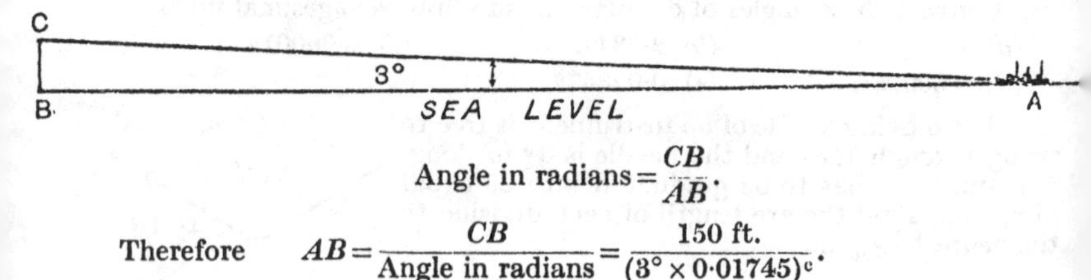

$$\text{Angle in radians} = \frac{CB}{AB}.$$

Therefore $$AB = \frac{CB}{\text{Angle in radians}} = \frac{150 \text{ ft.}}{(3° \times 0{\cdot}01745)^c}.$$

Therefore $$AB \text{ (in ft.)} = \frac{150}{0{\cdot}05235 \text{ ft.}}$$
$$= 2865 \text{ ft.} = 955 \text{ yd.}$$

7. A pilot is steering 090° C, but there is a steering error of $\frac{1}{2}$° and the plane flies on a course 089$\frac{1}{2}$° C. If the speed of the plane is 120 m.p.h., what distance is he to port of his intended position after 2$\frac{1}{2}$ hours' flight?

8. From the bridge of a destroyer 25 ft. above sea level a floating mine is observed. The reading on to it by sextant is 1$\frac{1}{2}$°. What is the range in yards?

9. A pilot is, by observation, 2 miles to starboard of his dead reckoning position after $\frac{3}{4}$ hour's flight at 80 m.p.h. What is his steering error in degrees and minutes?

10. A gun is sighted to 800 yd. range. The gunner is 10′ right of true alignment when he fires. How far wide will his shot be?

11. A gunner's shot hits a spot 4 yd. to the left of his target at 500 yd. range. How many minutes was his line of sight to the left of his objective?

PROBLEMS IN SIMPLE EQUATIONS

Hints for Solving Problems

1. Read the question *carefully*, several times, to be sure you clearly understand what is wanted.

2. Select from the question what you are asked to find, e.g. "Find the speed...", "What is the cost?" "How long will it take...?" This usually appears in the last part of the question.

3. Decide what unknown quantity you will take to be represented by a symbol. If there are two quantities to be found, such as two speeds in m.p.h., select a letter for the *smaller* speed and so avoid introducing unnecessary fractions of x.

4. Write a complete statement of your unknown quantity, and the unit you choose, e.g. "Let x be the speed of the slower plane in m.p.h.", or better, "Let x (m.p.h.) be the speed of the slower plane".

5. Find, if you can, two different ways of expressing the same thing. This gives you the equation you want.

6. Always check your answer by substitution in the original problem and *not* in any equation you may have made—you may have made a wrong equation.

EXAMPLE 1. The resistance of a wire is 5 ohms. The wire is divided into two parts such that the longer part offers 2 ohms more resistance than half of the resistance of the other part. Find the resistance of each part.

Let x ohms be the resistance of the *smaller* part. Then, since the resistance of the whole wire is 5 ohms, the resistance of the longer part is

$$(5-x) \text{ ohms.}$$

Half the resistance of the shorter part $=\frac{x}{2}$, and 2 ohms less than the longer part $=(5-x)-2$.

The problem tells us that these are equal, so the equation becomes:

$$\frac{x}{2}=(5-x)-2,$$

i.e.

$$\frac{x}{2}=3-x.$$

$\therefore\ x=6-2x,\ 3x=6$ and $x=2$.

Therefore the resistance of the shorter wire is 2 ohms.

The total resistance is 5 ohms, so that the resistance of the longer wire is 3 ohms.

EXAMPLE 2. What temperature in ° F. is exactly the same when expressed in ° C. (see Ex. VII, Question 9).

Let t be the required temperature. Then we know that

$$t^\circ \text{ F.}=t^\circ \text{ C.} \qquad \text{......(i)}$$

To express ° F. as ° C. we first subtract 32° and take $\frac{5}{9}$ of the answer.

So

$$t^\circ \text{ F.}=(t-32)\ \tfrac{5}{9}{}^\circ \text{ C.} \qquad \text{......(ii)}$$

From equations (i) and (ii) we get

$$t^\circ \text{ C.}=(t-32)\ \tfrac{5}{9}{}^\circ \text{ C.}$$

$$\therefore\ 9t=5t-160,$$

$$4t=-160,$$

i.e. $t=-40$ (or 40° below zero).

Therefore $-40°$ F. and $-40°$ C. are identical temperatures.

EXERCISE XLVII.

Solve the following problems:

1. What temperature in ° C. is exactly half of what it is when expressed in ° F.?

2. Two planes leave A for B at the same time. The first travels at 120 m.p.h. and arrives at B half an hour in front of the second plane. The second plane flies at $\frac{3}{4}$ of the speed of the first. What is the distance from A to B?

3. The total capacity of two petrol tanks is 500 gal. Their difference is 140 gal. Find the capacity of each tank.

4. One petrol tank holds 140 gal. more than another. When the large tank is half full and the small tank is full, there are 110 gal. more fuel in the small tank than in the large tank. What is the capacity of each tank?

5. One wire is twice as long as another. After 10 ft. have been cut from the longer wire and 10 ft. from the shorter wire, the longer wire is $2\frac{1}{2}$ times as long as the shorter one. Find the length of each wire.

6. If the radius of a particular circle is reduced by $3\frac{1}{2}$ in. its area is reduced by $192\frac{1}{2}$ sq. in. Find the radius of this circle (using $\pi = \frac{22}{7}$).

7. On a certain day the wind speed is 30 m.p.h. At what indicated air speed (I.A.S.) will the up-wind ground speed be $\frac{3}{5}$ of the down-wind ground speed?

8. If the air speed of a plane is increased by 10 % its speed is found to be 42 m.p.h. more than 80 % of the original air speed. What was the original air speed?

9. If two resistances only are placed in parallel the resultant resistance can be obtained directly as

$$\frac{\text{Product of the two resistances}}{\text{Sum of the two resistances}}.$$

Use this to solve the following:

What "shunt" (i.e. resistance in parallel) should be used with an ammeter, whose resistance is 200 ohms, so that the combined resistance of ammeter and shunt shall be 0·1 ohm. (Answer to four decimal places.)

10. Use the same method to solve this problem:

What condenser, if placed in series with another condenser of 5 microfarad capacity, will produce a combined capacity of $1\frac{1}{4}$ microfarads?

PROBLEMS IN SIMULTANEOUS EQUATIONS

EXAMPLE 1. The total capacity of two full tanks is 600 gal. When half of the contents of the larger tank has been used, and a quarter of the contents of the smaller, the tanks then hold equal quantities. What is the capacity of each tank?

Let x gal. be the capacity of the larger tank

and y gal. " " smaller tank.

Then total capacity is

$$x+y=600. \qquad \text{......(i)}$$

By taking $\frac{1}{2}$ the contents from the larger tank we have left $\frac{x}{2}$ gal.

By taking $\frac{1}{4}$ the contents from the smaller tank we have left $\frac{3y}{4}$ gal.

These amounts, we are told, are equal.

Therefore $\frac{x}{2}=\frac{3y}{4}$. ......(ii)

Cross multiplying, $4x=6y$,

or $4x-6y=0$. ......(iii)

Multiply (i) by 4 and subtract (iii), we get

$$\left.\begin{aligned}4x+4y&=2400\\4x-6y&=0\end{aligned}\right\}.$$

Thus $10y=2400$,

$y=240$ gal.

And, by subtraction $(600-240)$, $x=360$ gal.

The capacities of the tanks are:

$$\left.\begin{aligned}&\text{Larger tank}\quad 360\text{ gal.}\\&\text{Smaller tank}\quad 240\text{ gal.}\end{aligned}\right\}.$$

Example 2. The maximum speed of a plane A is 50 m.p.h. more than that of B. When A is flying at half its maximum speed it is 100 m.p.h. slower than the maximum speed of B. Find the maximum speeds of both planes.

Let · x (m.p.h.) be A's maximum speed

and y (m.p.h.) be B's " "

Then the difference in speeds is

$x-y=50$, ...(i)

and half A's maximum speed is $\frac{x}{2}$.

This time, B's maximum speed y is 100 m.p.h. faster than half A's maximum speed, i.e.

$y-\frac{x}{2}=100$. ......(ii)

Adding (i) and (ii),

$$x-\frac{x}{2}=150.$$

$$\therefore\ \frac{x}{2}=150.$$

and $x=300$ m.p.h.

And hence $y=250$ m.p.h.

Exercise XLVIII.

1. An aeroplane A travelling at maximum speed is flying twice as fast as plane B. Plane B then accelerates and increases speed by 60 m.p.h., but her speed is still only $\frac{2}{3}$ of A's maximum. Find the speed of A and B before B accelerated.

2. A plane flew 360 miles at a certain air speed down-wind and returned to base on a reciprocal course up-wind. The return journey occupied $1\frac{1}{2}$ hours longer and the ratio of return speed to outward speed was $\frac{2}{3}$. Find the wind speed and air speed of the plane.

3. Two petrol tanks are of unequal capacity. The small one is full and the large one half full. The large tank is filled from the smaller and 80 gal. remain in the small tank. When the large tank is full and the small one half full, and the small tank filled from the larger, then 320 gal. are left in the large tank. Find the capacity of each tank.

4. A rectangular plot was marked out on the ground. If the length had been increased by 10 yd. and the width reduced by 10 yd. the area would have been 2700 sq. yd. On the other hand, if the length had been reduced by 10 yd. and the width increased by 10 yd. the area would have been 3500 sq. yd. Find the length and width of the rectangle marked out.

5. The difference between two numbers is 3 and the difference between their squares is 39. Find the two numbers. (Use factors to solve this.)

6. An increase of altitude of $12\frac{1}{2}\%$ will enable a plane to clear an approaching range of hills by 300 ft. If instead the altitude is increased by 200 ft., the clearance will be only 240 ft. Find the present altitude of the plane above sea level and the height of the range of hills.

7. At a certain altitude the pressure of the air in millibars was half that at sea level. At a further 3000 ft. altitude the pressure was 405 mb. Find the pressure at sea level and the altitude at which the pressure was half that at sea level. (Pressure decreases 1 mb. for each 30 ft. rise in altitude.)

8. A condenser plate has an area A sq. cm. and perimeter 22 cm. If the length is reduced by 2 cm. and the breadth increased by 1 cm., the area remains the same. Find the dimensions of the rectangular plate.

9. The joint resistance of two wires in parallel is 2 ohms. If the resistance of the first wire is halved and that of the second doubled, the joint resistance is still 2 ohms. Find the resistance of each wire.

10. A long cable AB, whose resistance, when perfect, was 400 ohms, sprang a leak to earth in an unknown place. The resistance of the leak itself was also unknown. When the end B was insulated a resistance test from A through the leak gave 300 ohms. When the end A was insulated a resistance test from B through the leak gave 600 ohms. Find the resistance at the leak and the resistance between A and the leak.

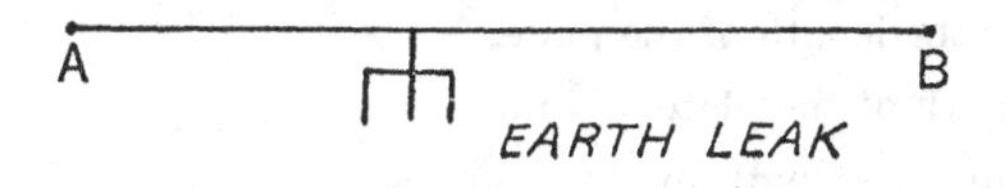

PROBLEMS IN QUADRATIC EQUATIONS

EXAMPLE 1. A plane with air speed 120 m.p.h. flies 6 miles up-wind. Speed is then increased to air speed 150 m.p.h. for another 6 miles up-wind. The first 6 miles take 1 minute longer to fly than the second 6 miles. What is the speed of the wind?

Let x (m.p.h.) be the wind speed. Then the ground speed for

the first 6 miles $=(120-x)$ m.p.h. ......(i)

and for the second 6 miles $=(150-x)$ m.p.h. ......(ii)

At ground speed (i) the plane flies $(120-x)$ miles in 60 min., i.e.

$$1 \text{ mile in } \frac{60}{(120-x)} \text{ min.}$$

or

$$6 \text{ miles in } \frac{6\times 60}{(120-x)} \text{ min.}$$

And at ground speed (ii) the plane flies

$$6 \text{ miles in } \frac{6\times 60}{(150-x)} \text{ min.}$$

The difference between these two times is given as 1 min.

$$\therefore \frac{6\times 60}{(120-x)} - \frac{6\times 60}{(150-x)} = 1.$$

Simplifying, $360(150-x)-360(120-x)=(120-x)(150-x)$.

Clearing brackets,

$$54000-\cancel{360x}-43200+\cancel{360x}=18000-270x+x^2.$$

$$\therefore x^2-270x+7200=0.$$

Factorising, $(x-240)(x-30)=0.$

$$\therefore x=240 \text{ m.p.h. or } 30 \text{ m.p.h.}$$

The value $x=240$ m.p.h. would mean that the plane was flying backwards and is consequently inadmissible (although mathematically correct), so the correct answer is

Wind speed $=30$ m.p.h.

EXAMPLE 2. The rectangular plate of a condenser has an area of 24 sq. in. If the length is decreased by 2 in. and the breadth increased by 1 in. the area is unaltered. What are the dimensions of the plate?

Let x in. $=$ the length of the plate.

Then breadth of the plate $=\dfrac{24}{x}$ in.

By reducing the length by 2 in., we have

$(x-2)$ in. for the new length.

And increasing breadth by 1 in., we have

$\left(\dfrac{24}{x}+1\right)$ in. for the new breadth.

The area is still 24 sq. in.

So that $$(x-2)\left(\frac{24}{x}+1\right)=24,$$

i.e. $$(x-2)\frac{24+x}{x}=24.$$

$$\therefore (x-2)(24+x)=24x.$$

Clearing brackets, $$24x-48+x^2-2x=24x.$$

Equating to zero, $$x^2-2x-48=0.$$

Factorising, $$(x-8)(x+6)=0.$$

Thus $$x=8 \text{ or } -6.$$

The only answer of value is $x=8$, since a plate with side -6 in. long means nothing.

Thus if $x=8$, the length of the plate is 8 in. and the breadth is $\frac{24}{8}=3$ in.

EXERCISE XLIX.

1. If a number is increased by 2 the value of its reciprocal is reduced by $\frac{1}{12}$. Find the number.

2. A plane flew 320 miles at a certain speed and then reduced speed by 40 m.p.h. and continued flying on the same course. The pilot landed after a total flying time of 4 hours, having covered in all 560 miles. Find his first speed and state why one of your answers is useless.

3. A pilot covered a distance of 360 miles at a certain speed. Had he flown at 60 m.p.h. less his journey would have taken him 1 hour longer. Find his speed.

4. Two condensers have square plates. The difference in the areas of these two plates is 24 sq. cm. If the side of the larger plate is 2 cm. longer than the side of the smaller plate, what are the actual lengths of one side of each plate?

5. One condenser plate is square and another rectangular, but their areas are the same. The long side of the rectangle is 3 cm. longer than a side of the square and the short side is 2 cm. shorter than a side of the square. What is the area of each plate?

6. The square of a certain number exceeds the sum of the squares of its two halves by 32. What is the number?

7. A certain quantity exceeds its reciprocal by $\frac{5}{6}$. Find this quantity.

8. A rectangle marked on the ground has an area of 250 sq. yd. Its length is $2\frac{1}{2}$ times its width. What are its dimensions?

9. A plane travelling up-wind with air speed 90 m.p.h. timed a 4 mile run and then after increasing air speed to 120 m.p.h. timed a second 4 mile run on the same course. The difference in time was $1\frac{1}{3}$ min. Find the wind speed.

10. The same plane performed the same operations, but this time down-wind. The difference in time was 24 sec. What was the wind speed?

CONSTRUCTION AND USE OF FORMULAE

EXAMPLE 1. Let

V represent the speed of a plane in m.p.h. in still air,
W ,, a wind speed in m.p.h.,
D ,, the distance flown in miles,
t ,, time taken in hours.

Then with the wind astern the speed of the plane is increased to

$$(V+W) \text{ m.p.h.}$$

But we know that

Distance flown = Speed × Time taken.

Thus $$D=(V+W)\,t.$$

This formula is suitable for calculating the distance, in miles, flown in any number of hours for any particular speeds of plane or wind.

Suppose the "still air" speed of plane = 110 m.p.h. Wind speed = 30 m.p.h. (astern). What distance is covered in $1\frac{1}{2}$ hours?

$$\begin{aligned} D &= t(V+W) \\ &= \tfrac{3}{2}(110+30) = \tfrac{3}{2}(140) \\ &= 210 \text{ miles.} \end{aligned}$$

Simple transposition will give us any other symbol, should that one be the unknown, i.e.

$$W=\left(\frac{D}{t}-V\right) \text{ m.p.h. and } t=\frac{D}{(V+W)}.$$

EXAMPLE 2. "Wind-finding" is the determination of the speed of the wind, and one method employed in wind-finding is as follows.

Selecting a prominent object, a pilot flies over it, notes the time, and with constant air speed, flies in a down-wind direction (i.e. with wind astern) for 3 min. Turning rapidly on to a reciprocal *true* course he flies back with the same air speed for a further 3 min. and then takes note of the time taken to return to the object.

Suppose the time thus taken is t min. Then let A = plane's air speed in m.p.h. and W = wind speed in m.p.h.

The pilot's down-wind speed $=(A+W)$ m.p.h.
,, ,, up-wind ,, $=(A-W)$ m.p.h.

And the distance flown down-wind in 3 min. $=\dfrac{A+W}{20}$ miles.

The distance flown up-wind in 3 min. $=\dfrac{A-W}{20}$ miles.

Therefore, after 6 min., the pilot is the *difference* of these two distances from his object, i.e.

$$\frac{A+W}{20}-\frac{A-W}{20} \text{ miles}$$

$$=\frac{A+W-A+W}{20}=\frac{2W}{20}=\frac{W}{10} \text{ miles.}$$

The time taken to reach his object from this point is t min.

His speed during this part of the flight is $(A-W)$ m.p.h.

Therefore $$t=\frac{\frac{W}{10}\times 60}{A-W}=\frac{6W}{A-W},$$

or $$6W=t(A-W)=At-Wt.$$

$$\therefore\ 6W+Wt=At \quad \text{and} \quad W(t+6)=At.$$

Therefore $$W \text{ (wind speed)}=A\left(\frac{t}{t+6}\right) \text{ m.p.h.}$$

EXERCISE L.

1. Following the same steps as in Example 2, work out a formula to find W in m.p.h. when the time down-wind and up-wind is not 3 min. but x min. in each case.

2. Use $x=4$ in your formula so obtained and check the wind speeds in the 2 min. column in the following table. Work to the nearest m.p.h. Then fill in the calculated wind speeds when $t=4$ and $t=6$ (in each case after 4 min. down- and 4 min. up-wind flying).

	t min., i.e time to reach object after 4 min. down- and 4 min. up-wind flying		
Air speed	$t=2$ min.	$t=4$ min.	$t=6$ min.
	W	W	W
100 m.p.h.	20 m.p.h.		
110 m.p.h.	22 m.p.h.		
120 m.p.h.	24 m.p.h.		

3. A plane has a speed of V m.p.h. Express the distance, D miles, travelled in (*a*) T hours, (*b*) 5 min., (*c*) x min.

4. The tank capacity of a plane is G gal. The reserve is x %. Express the usable fuel, P gal., in terms of G and x.

5. A plane travelled x miles in y min. What was the speed, V, in m.p.h.?

6. In what time, T minutes, will a plane travel x miles if its speed is M m.p.h.?

7. A plane has to cover z miles. How far will it be from its objective after flying for x min. at y m.p.h.?

8. A plane uses P lb. of fuel per H.P. hour and 1 gal. of petrol weighs 7·2 lb. Find G, the fuel used in gallons, in T hours, if the plane has twin engines each of x H.P.

9. A plane travels for x hours at y m.p.h. and then for a hours at b m.p.h. What is A, the plane's average speed?

10. During a flight of total time y hours, a plane flies at two altitudes. For the first part, x hours, the plane flies at b ft. and for the rest of the journey at c ft. The average altitude during the flight is a ft. Express the time, in hours, at altitude c ft., in terms of the remaining symbols.

11. The endurance of a plane at any air speed is the number of hours it can remain in the air at that speed without using its fuel reserve.

At an air speed of V m.p.h. a plane, with total tank capacity T gal., uses x gal. of petrol per hour. Make a formula for E (petrol-hours of endurance) at this speed, allowing for a reserve of $y\%$ of the tank capacity.

12. The "radius of action" of a plane is the maximum distance it may fly on its outward course and still be able to return without drawing on its fuel reserve.

Let D = radius of action (distance out) in miles,
" V_1 = ground speed out at air speed A m.p.h.,
" V_2 = ground speed home at air speed A m.p.h.

Find (*a*) a formula for total time of flight.

Now let y be the number of gallons of fuel used per hour at *air speed* A m.p.h. and G the usable petrol in gallons.

Find (*b*) a formula for the total time of flight based on petrol consumption.

Find (*c*), by combining (*a*) and (*b*), another formula for radius of action (D) in terms of the other quantities.

13. Make a diagram to help with the following:

A mooring buoy is anchored to sea bottom at a spot charted with depth of water x fathoms at low water spring tide (L.W.S.T.). The mooring chain from buoy to anchor is y fathoms in length. Express, in terms of x and y, the radius of swing of the buoy in yards at (*a*) L.W.S.T., (*b*) half tide on a day when full spring tide rises z feet above L.W.S.T.

(1 fathom = 6 ft.).

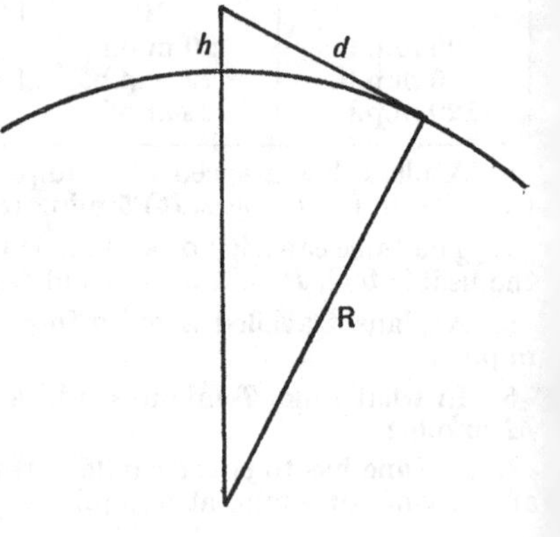

14. The figure shows a section of the earth. Find a formula for d, which is the visibility distance in miles from a point h miles above ground. R is the radius of the earth in miles.

RELATIVE VELOCITY

Some attention has already been given to this subject in Part I, under the section on relative speeds.

Its application has, however, been very limited since no attention has yet been given to the consideration of any speeds which are not in the same or opposite directions.

It is obvious that conditions involving the relative movement of objects with quite different velocities are far more common in actual practice and the solution of such problems calls for an extension of the theory of the triangle of velocities.

The relative velocity of a moving body *B*, to another moving body *A*, is the speed and direction which *B* appears to have to an observer on *A*.

In fig. 21, *A* represents the position of a ship which is steering a course 028° T at 10 knots. The point *B* represents the position of another vessel distant 10 N.M. from *A*, and on a true bearing 105° from *A*.

Vessel *B* is steering a course 330° T at 12 knots. Since the observer on *A* is himself in motion, the movement of *B* does not appear to him to be along *BC*. He can appreciate only the *relative* motion of *B*, and to determine this it is best to imagine *A* to be brought to rest.

Suppose *A* were steaming against a head tide of 10 knots. Then *A* would be actually at rest relative to sea bottom.

(NOTE. An observer can always be brought to rest by giving him an equal and opposite velocity to the one he already possesses.)

This head tide of 10 knots would also affect *B*, so we must apply to *B* the same velocity (speed and course) as would bring *A* to rest, i.e. *A*'s velocity reversed.

This velocity in fig. 21 is represented by vector *BD*, where *BD* is parallel to *AF*. The ship *B* has now two velocities:

(*a*) Its own. 12 knots at course 330° T (vector *BC*).

(*b*) The imaginary tide of 10 knots setting at 208° T (vector *BD*).

By the parallelogram of velocities the resultant of these two is the vector *BE*, and this is the path along which *B* appears to be travelling to the observer on *A*.

Vessel *B*, therefore, appears to travel on a course 278° T at 10·9 knots. This is called the *relative velocity* of *B* with respect to *A*.

Suppose now we produce *BE*, the vector of *B*'s relative track. We find that the relative track of *B* passes astern of *A* and is at its nearest point to *A* by the perpendicular distance *AK*. (Notice that *AK* is less than *AO*, where *O* is the point at which *B* actually crosses the track of *A*.)

If we wish to know at what time B is nearest to A we measure BK (= 9·9 N.M. on the distance scale) and calculate how long B will take to travel this 9·9 N.M. at 10·9 knots.

Time taken is $\frac{9\cdot9}{10\cdot9} \times 60$ min. = 54·4 min.

Consider the Red or Green (fore and aft line) bearing of A to B.

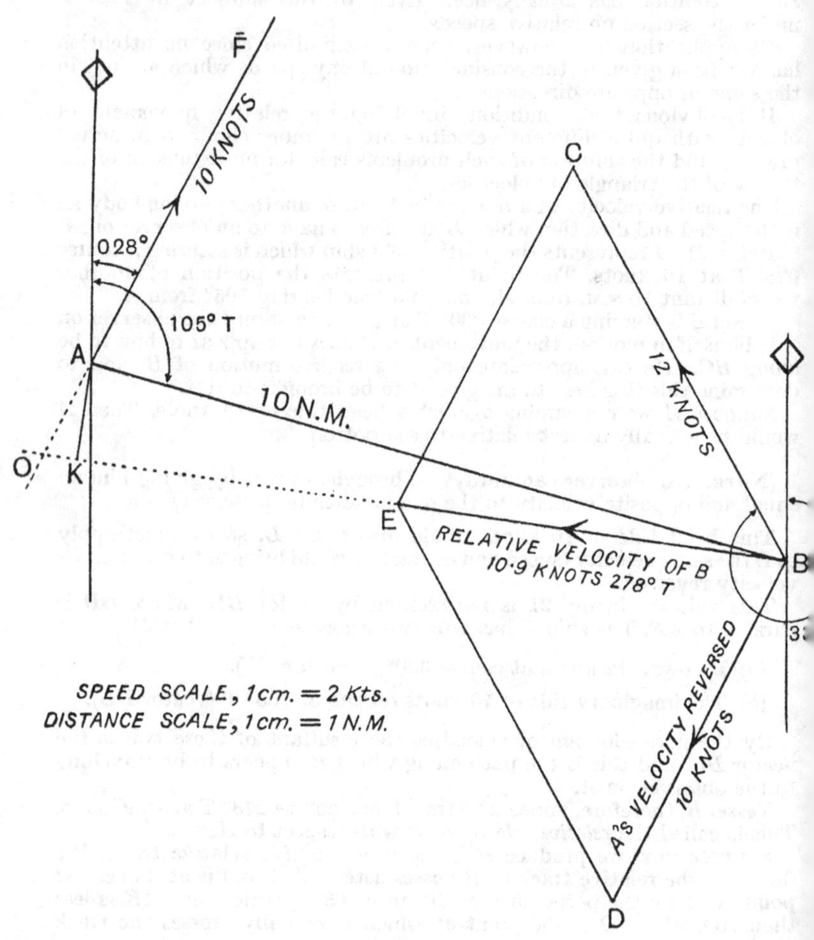

Fig. 21

While at B this is Green 077°. As B moves along track BKO this bearing increases up to 180° at O and then changes to Red and decreases.

In these problems it should be noted that the distance scale can be quite independent of the speed scale and, to save space and paper, a smaller scale may be used. Always be careful to measure the distance by the distance scale and the speed by the speed scale.

In fig. 22 it happens that the relative velocity of B to A coincides with the line of sight BA.

In other words the bearing of B from A relative to fore and aft line, or by compass, will remain constant throughout, i.e. Green 050° or True 097°.

Under these conditions the ships must collide at point O.

This is one of the methods used to check the possibility of collision

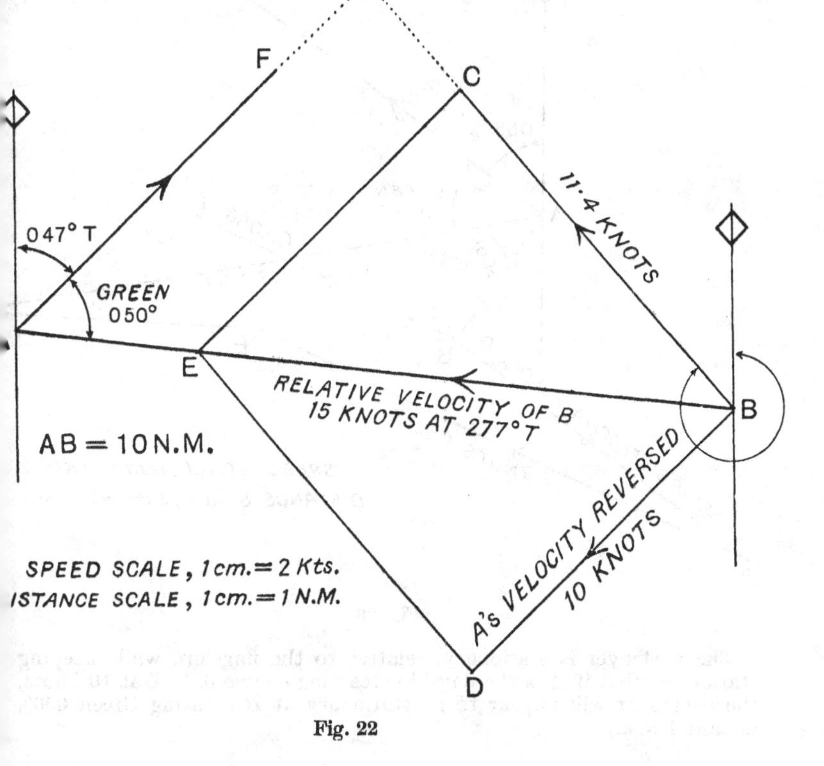

Fig. 22

at sea. The observer on *A* takes frequent bearings on *B*. If the bearing does not change then *A* must alter course.

The time of collision would be time for *B* to reach *A*, i.e. time to travel 10 N.M. at the relative speed 15 knots.

Another interesting example of relative velocity is to be found in ships changing station.

A flagship is steaming at 10 knots on a course 055° T and is accompanied by a destroyer on the same course and speed, keeping station distant 1 N.M. on the starboard beam of the flagship (i.e. bearing Green 090° from flagship). The destroyer is now ordered to change station and take up position 4 N.M. on the starboard bow (i.e. bearing Green 045° from the flagship), using 15 knots while changing station.

Fig. 23 shows how we may find the course for her to steer, by using relative velocity.

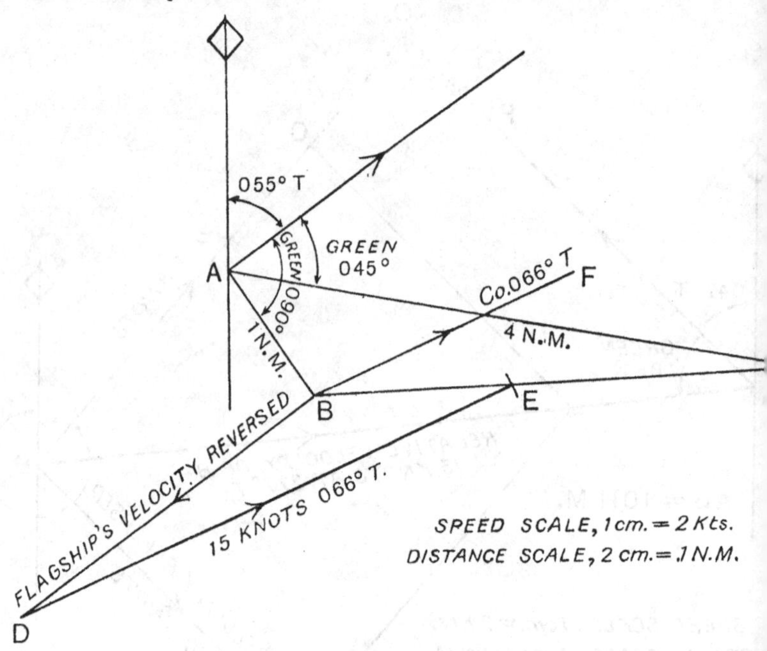

Fig. 23

The destroyer is stationary, relative to the flagship, while keeping station, so that if *A* is the flagship steaming course 055° T at 10 knots, the destroyer will appear to be stationary at *B* (bearing Green 090°, distant 1 N.M.).

The new position that the destroyer has to take up is C, 4 N.M. from A on the starboard bow (Green 045°). Therefore BC is the distance the destroyer must make good to reach C. BC is 6·8 cm. long = 3·4 N.M.

Immediately the destroyer alters course she is no longer stationary in respect to the flagship, so now we must find her relative velocity, as we did in the example in fig. 21.

At B lay off BD, the vector of the reverse of the flagship's velocity (235° T at 10 knots).

The destroyer's new speed is 15 knots. With centre D and compass open to 7·5 cm. (15 knots of speed scale) mark off E where the arc cuts the track BC.

Then DE is the destroyer's true course (=066° T). As, in our figure, it starts from B and not from D, transfer it by parallels to BF.

BE (2·8 cm.), representing 5·6 knots, is the new speed of the destroyer, *relative* to the flagship.

The time to get to the new station will therefore be the time to make good 3·4 N.M. at 5·6 knots $= \frac{3{\cdot}4}{5{\cdot}6} \times 60$ min. $= 36{\cdot}8$ min.

EXERCISE LI.

Find the relative velocity of B with respect to A in the following and state the nearest approach, passing ahead or astern:

	Velocity of A	Bearing and distance of B from A	Velocity of B
1.	060° T 10 knots	5 N.M. 090° T	350° T 10 knots
2.	315° T 12 knots	6 N.M. Green 120°	035° T 10 knots
3.	000° T 15 knots	6 N.M. 080° T	220° T 10 knots
4.	175° T 11 knots	10 N.M. Red 132°	130° T 14 knots
5.	265° T 12 knots	8 N.M. 062° T	220° T 12 knots

6. A seaplane carrier is steaming on a course 080° T at 15 knots. A destroyer keeping station distant 5 cables (i.e. 0·5 N.M.) on the starboard bow of the carrier (i.e. Green 045°) is ordered to take station distant 10 cables on the carrier's starboard beam, using 20 knots. Find

(*a*) Destroyer's True Course.

(*b*) Distance to be made good.

(*c*) Time to get to new station.

7. If the destroyer in Question 6 is later ordered to return to her former station, find again (*a*), (*b*) and (*c*), the same three factors concerned.

8. A squadron of 5 planes is flying in V formation on a course 030° T at A/S 80 m.p.h. Two planes are on a line of bearing Green 135° from the leader and the remaining two on a line of bearing Red 135°. They are

keeping station at intervals of 1 cable (200 yd.). The rear starboard plane is ordered to scout to a position 20 miles Green 070° from leader and return using A/S 120 m.p.h. Neglect wind velocity. Find (*a*) outward speed made good, (*b*) outward course, (*c*) outward time, (*d*) inward speed made good, (*e*) inward course, (*f*) inward time.

9. Two planes *A* and *B* are on a course 010° T at A/S 80 m.p.h. Plane *A* bears Green 040° 1 mile from *B*. The leading plane is ordered to open her distance from the rear plane to 10 miles, preserving the bearing during flight and using 100 m.p.h. and then to return to former position. Find (*a*) outward course, (*b*) outward speed made good, (*c*) outward time, (*d*) inward course, (*e*) inward speed made good, (*f*) inward time.

10. At time 1200 a plane *A* is flying at altitude 5000 ft. with A/S 90 m.p.h. on a course 035° T. Another plane *B* is sighted bearing 095° T, distant 5 miles, flying at the same altitude with A/S 100 m.p.h. on a course 300° T. Find

(*a*) Time of nearest approach (to nearest min.).
(*b*) Time when *B* crosses track of *A* (first decimal place).
(*c*) Time when *B* bears Red 160° (first decimal place).
(*d*) Distance of nearest approach.

11. A mooring buoy for a seaplane is laid to southward of a coast which runs S.W. to N.E. The buoy is 2·5 N.M. from the nearest point on the coast. If a seaplane, which is moored to it, breaks her moorings when the tide is setting 010° T at 4 knots, how long will it be before she runs aground?

12. A seaplane alights in harbour 6 cables from her mooring buoy and finds the buoy to bear 068° T. The tide is setting 345° T at 3 knots. The seaplane taxis at 6 knots to reach the buoy. Find

(*a*) The course to steer, allowing for tide.
(*b*) The speed of approach.
(*c*) Time to reach the buoy.

LOGARITHMS

In the section on signs and symbols we have seen that it is convenient to write

$$a \times a \times a \times a \times a \quad \text{as} \quad a^5,$$

where the small figure 5 is called the index, or power, to which a is raised.

We have had, also, numerous examples of division by the subtraction of indices, such as

$$a^5 \div a^2 = a^{5-2} = a^3.$$

We may do the same thing with numbers.

Thus $10{,}000 = 10 \times 10 \times 10 \times 10 = 10^4$.

Also $\frac{10{,}000}{100} = \frac{10^4}{10^2} = 10^{4-2} = 10^2 = 100$.

In words we say that 100 is the "base" 10 raised to the power 2, and 1000 is the "base" 10 raised to the power 3.

Another way of expressing this is to say that the *logarithm* of 100 to base 10 is 2, and the *logarithm* of 1000 to base 10 is 3.

The logarithm of any number between 100 and 1000 will be between 2 and 3. It is expressed as a decimal.

The logarithm of 200 to base 10 is approximately 2·3.

This may be shown in this way:

It is found by multiplication that 2^{10} is 1024 which is a little more than 1000, i.e. 10^3.

So that $2^{10} = 10^3$ (approximately).

And
$$2^{10} \times 100^{10} = 10^3 \times 100^{10}$$
$$= 10^3 \times 10^{20} = 10^{23} \text{ (approx.)}.$$

Therefore $200^{10} = 10^{23}$ (approx.).

And $200 = 10^{2\cdot 3}$ (approx.).

So that the logarithm of 200 to base 10 is nearly 2·3. We can get closer approximation by using more decimal places; four are usually sufficient.

Since all logarithms which we use are to the base 10, we call them simply 'logarithms'.

Thus we write log 100 = 2, log 1000 = 3, and log 200 = 2·3010.

All logarithms consist of two parts:

(*a*) The whole number part, called the *Characteristic*.

(*b*) The decimal part, called the *Mantissa*.

The mantissa is always obtained from a table of logarithms, and one will be found at the end of the book which will give the logarithm of any number provided it consists of not more than four digits, i.e. 0–9999.

The characteristic, or whole number part, of the logarithm is obtained by applying two very simple rules, which are best explained and most easily mastered by studying examples.

Turn to the table of "Logarithms of Numbers". These tables are called "four-figure logarithms". This means that we cannot obtain from them the logs of numbers containing more than four figures except under special conditions, and also that the mantissa, or decimal portion, is given in the table to four decimal places only.

Let us find the logarithm of 2146, i.e. log 2146. The first two figures of the number, i.e. 21, are found in the left-hand column of the table. The third figure of the number is a 4, so we travel horizontally along the 21 line until we are under number 4 at the top of the table. This gives us the figures 3304. This is the mantissa portion of log 214. Similarly, under the 5 column, 3324 would be the mantissa portion of log 215.

We are finding log 2146, so we look for the last fig. 6 in the columns on the right-hand side (called the "difference" columns). We have seen that the mantissa increases as the number is greater, so the differences are *added*. Along the 21 line, under the 6 in the difference column, we find the difference to be added is 12.

Therefore the mantissa of log 2146 = 3304 + 12 = 3316.

To obtain the characteristic, or whole number portion of the log, the first rule is

(i) Count the number of figures in the whole number of the quantity in question.

Subtract 1 to obtain the characteristic. Thus 2146 is a whole number of 4 digits.

Therefore the characteristic of its logarithm is 3.

Notice if the number we were using had been 214·6 the whole number portion of this is 3 digits, so the characteristic would be 2.

The full logarithm of 2146 is, therefore,

$$\log 2146 = 3{\cdot}3316.$$

Rule (ii) will be explained later.

Suppose now we want log 21·46 instead of log 2146.

We take no notice of the decimal point, but refer to the tables exactly as before to find the mantissa 3316.

This time the number of digits in the whole number part of 21·46 is 2, so the characteristic is 1, and

$$\log 21{\cdot}46 = 1{\cdot}3316.$$

Exercise LII.

1. Check the following from the tables:

Number	Mantissa for first 3 figures	Added difference	Full mantissa	Characteristic
1122	0·0492	8	0·0500	3
1554	0·1903	11	0·1914	3
30·30	0·4814	0	0·4814	1
5·678	0·7536	6	0·7542	0

2. From the tables find the log of each of the following:

(*a*) 212·4. (*b*) 346·8. (*c*) 34·68. (*d*) 3468.
(*e*) 81·79. (*f*) 7·642. (*g*) 764·2. (*h*) 220·7.
(*i*) 2200. (*j*) 23. (*k*) 3. (*l*) 2.
(*m*) 200. (*n*) 3000. (*o*) 4001. (*p*) 400·1.
(*q*) 2·148. (*r*) 3·005. (*s*) 73250. (*t*) 897000.

Antilogarithms.

The accuracy of the answers obtained in Exercise LII may be checked by what are called antilogarithms. This merely involves working from the tables.

EXERCISE LIII.

Use antilogs for all your answers in Exercise LII and see that the answers so obtained are identical with the numbers given in the respective questions.

Logarithms of decimals without whole numbers.

We know that log 31·52 has characteristic 1 and log 3·152 has characteristic 0.

By the same rule (one less than the digits of the whole number portion) log 0·3152 will have characteristic −1.

This is usually written

$$\log 0{\cdot}3152 = \bar{1}{\cdot}4986,$$

to indicate that the characteristic *only* is negative. The mantissa is always positive.

If we divide 31·52 by 10, we get 3·152, and have reduced the characteristic of the log by 1.

If we divide 3·152 by 10, we get 0·3152, and have again reduced the characteristic of the log by 1.

If we divide 0·3152 by 10, we get 0·03152, and again we reduce the characteristic of the logarithm by 1.

But the $\log 0{\cdot}3152 = \bar{1}{\cdot}4986$.

Therefore $\log 0{\cdot}03152 = \bar{2}{\cdot}4986$, and consequently

$$\log 0{\cdot}003152 = \bar{3}{\cdot}4986.$$

Thus, the characteristic of a decimal is always negative and is 1 more than the number of 0's immediately after the decimal point. This is Rule (ii).

EXERCISE LIV.

Find the logarithms of the following and check your answers by antilogs:

1. 0·0614.	2. 0·3213.	3. 0·0008.	4. 0·008222.
5. 0·81.	6. 0·000005.	7. 0·03021.	8. 0·00762.
9. 7·62.	10. 0·762.	11. 0·0777.	12. 7·77.
13. 79·1.	14. 0·0003.	15. 303·2.	16. 0·000041.
17. 0·83.	18. 0·021.	19. 2·6.	20. 0·0053.

Uses of logarithms.

(*a*) *Multiplication.* If we wish to multiply two numbers we merely add their logs (as we add indices of like symbols in algebra) and then use antilogs to get the answer.

EXAMPLE 1. Thus $7{\cdot}8 \times 3{\cdot}4$.

Taking logs, $\log 7{\cdot}8 + \log 3{\cdot}4 = 0{\cdot}8921 + 0{\cdot}5315$

$= 1{\cdot}4236.$

By antilogs, the answer is 26·52.

EXAMPLE 2. Simplify $0{\cdot}07841 \times 292{\cdot}6$.

Then $\log 0{\cdot}07841 + \log 292{\cdot}6 = \bar{2}{\cdot}8944 + 2{\cdot}4664$
$= 1{\cdot}3608.$

By antilogs, the answer is 22·95.

If more than two quantities are involved in the multiplication, add all the logs and then use antilogs for the answer.

EXERCISE LV.

Work the following by logarithms.

1. $20{\cdot}1 \times 196{\cdot}7$.
2. $54{\cdot}41 \times 37{\cdot}13$.
3. $2{\cdot}3 \times 4{\cdot}92 \times 31{\cdot}63 \times 0{\cdot}009$.
4. $246{\cdot}8 \times 0{\cdot}08 \times 0{\cdot}0212 \times 1{\cdot}834$.
5. $45{\cdot}2 \times 0{\cdot}913 \times 0{\cdot}05 \times 1{\cdot}212$.
6. $0{\cdot}007 \times 0{\cdot}02 \times 0{\cdot}006$.
7. $23{\cdot}74 \times 0{\cdot}0005 \times 2{\cdot}621 \times 3{\cdot}753$.
8. $21{\cdot}21 \times 0{\cdot}301 \times 205{\cdot}6 \times 0{\cdot}00084$.
9. Find the value of $3{\cdot}142 \times R \times R$ when $R = 2{\cdot}3$.
10. Find the value of $2 \times 3{\cdot}14 \times R$ when $R = 7{\cdot}314$.

(*b*) *Division.* Just as multiplication is addition of indices, so is division the subtraction of indices.

EXAMPLE 1. Simplify $13{\cdot}72 \div 2{\cdot}981$.

By logs, $\log 13{\cdot}72 - \log 2{\cdot}981 = 1{\cdot}1373 - 0{\cdot}4743$
$= 0{\cdot}6630.$

By antilogs, the answer is 4·602.

Be careful of your subtraction when you have negative characteristics.

EXAMPLE 2. $2{\cdot}087 \div 0{\cdot}4161$.

By logs, $\log 2{\cdot}087 - \log 0{\cdot}4161 = 0{\cdot}3196 - \bar{1}{\cdot}6192.$
$= 0{\cdot}3196 + 1 - 0{\cdot}6192$
$= 1{\cdot}3196 - 0{\cdot}6192 = 0{\cdot}7004.$

By antilogs, the answer is 5·017.

EXAMPLE 3. Simplify $1 \div 5{\cdot}871$.

First of all, what is log 1?

We know $\log 100 = 2$ (or 2·0000), i.e. characteristic 1 less than number of digits; log 10 is 1·0000, or characteristic is again 1 less than number of digits; so that log 1 will be 0·0000, with characteristic again 1 less than number of digits.

Another way to prove that $\log 1 = 0$ is to write $\frac{100}{100} = 1$, i.e.

$$\frac{10^2}{10^2} = 1 \quad \text{or} \quad 10^{2-2} = 1.$$

$$\therefore\ 10^0 = 1, \text{ so that } \log 1 = 0.$$

Therefore taking logs of $1 \div 5{\cdot}871$, we have

$$\log 1 - \log 5{\cdot}871 = 0{\cdot}0000 - 0{\cdot}7687.$$

By subtraction, $= \bar{1}{\cdot}2313.$

By antilogs, the answer is 0·1704.

EXERCISE LVI.

Work the following divisions by logarithms:

1. $\frac{14{\cdot}61}{2{\cdot}306}$.   2. $\frac{2041}{13{\cdot}5}$.   3. $\frac{2{\cdot}4}{0{\cdot}0072}$.   4. $\frac{0{\cdot}061}{0{\cdot}0824}$.

5. $\frac{0{\cdot}08}{712{\cdot}3}$.   6. $\frac{0{\cdot}014}{31{\cdot}23}$.   7. $\frac{0{\cdot}005}{2{\cdot}2}$.   8. $\frac{0{\cdot}31}{0{\cdot}0031}$.

9. $\frac{0{\cdot}063}{6{\cdot}3}$.   10. $\frac{3{\cdot}178}{6{\cdot}245}$.

EXAMPLE 4. With mixed multiplication and division it is better to work numerator and denominator separately.

Simplify $\frac{3{\cdot}81 \times 12{\cdot}2 \times 7{\cdot}805 \times 0{\cdot}137}{4{\cdot}52 \times 3{\cdot}6}$.

Numerator

log 3·81	=0·5809
log 12·2	=1·0864
log 7·805	=0·8924
log 0·137	$=\bar{1}{\cdot}1367$
log numerator	=1·6964
log denominator	=1·2114
log answer	=0·4850

Denominator

log 4·52	=0·6551
log 3·6	=0·5563
log denominator	=1·2114

By antilogs, the answer is 3·055.

EXERCISE LVII.

1. Simplify $\frac{20{\cdot}3 \times 1{\cdot}781 \times 3{\cdot}005}{63{\cdot}25 \times 0{\cdot}99}$.

2. Simplify $\frac{209 \times 3{\cdot}26 \times 0{\cdot}08}{4{\cdot}7 \times 8{\cdot}213}$.

3. Convert 10,000 ft. to metres, if 1 metre = 3·26 ft.

4. Change 96 N.M. to statute miles, if 1 N.M. = $\frac{76}{66}$ miles.

5. Convert 162 Km. to N.M., if 1 Km. = $\frac{41}{76}$ N.M.

6. Find A from the formula $A=\frac{L \times S}{R}$, if $L=2150$, $S=0{\cdot}00000067$, $R=5{\cdot}2$.

7. Find the distance flown in 1 hr. 10 min. if the speed is $2\frac{1}{2}$ miles in 65 sec.

8. Find the distance flown in kilometres in 1 hr. 48 min. if the speed is 3 Km. in 37 sec.

9. Change a speed of $1\frac{1}{2}$ miles in 51 sec. to m.p.h.

10. Find the time taken for a journey of 217 N.M. if the speed is 147 S.M. per hour.

Squares, square roots, cubes and cube roots by logarithms.

As we have seen, the logarithm of the *product* of two or more numbers is the *sum* of their logarithms.

Consider $a \times a \times a$, where a is any number.

The logarithm of this is $\log a + \log a + \log a = 3 \log a$.

Thus $$\log a^3 = 3 \log a.$$

Similarly $$\log a^2 = 2 \log a.$$

Therefore, to find the log of the square, or cube, of a number, we multiply its log by 2, or 3, respectively.

Conversely, to find the log of the square root, or cube root, of a number we divide its log by 2, or 3, respectively.

This is much the easiest way to find square, or cube, or higher roots of numbers.

Exercise LVIII.

Find the value of:

1. $\sqrt{109{\cdot}7}$.
2. $\sqrt[3]{84{\cdot}31}$.
3. $(7{\cdot}532)^2$.
4. $(2{\cdot}18)^3$.
5. $(1{\cdot}917)^4$.
6. $\dfrac{(3{\cdot}016)^3}{(2{\cdot}14)^2}$.
7. $\dfrac{(1{\cdot}914)^2}{(3{\cdot}6)^3}$.
8. $\sqrt{(2{\cdot}4)^3}$.
9. $\sqrt[3]{(4{\cdot}73)^2}$.
10. $\sqrt{16{\cdot}81} \times \sqrt[3]{(3{\cdot}5)^2}$.

Exercise LIX. Problems involving logarithms (assume $\pi = 3{\cdot}142$).

1. What is the area of a circle of radius 1·7 in.?

2. The area of a sphere is $4\pi R^2$. What is the area when $R = 2{\cdot}316$ in.?

3. The volume of a sphere is $\frac{4}{3}\pi R^3$. What is the volume if $R = 1{\cdot}726$ ft.?

4. The volume of a cylinder is $\pi R^2 H$ (where H is the height of the cylinder). What is the volume if $R = 2{\cdot}31$ in. and $H = 4{\cdot}415$ in.?

5. The volume of a cone is $\frac{1}{3}\pi R^2 H$ (where H is the height of the cone). What is the volume if $R = 1{\cdot}75$ in. and $H = 3{\cdot}5$ in.?

6. The area of the curved surface of a cone is πRh, where h is the slant height. What is the area if $R = 3{\cdot}142$ cm. and $h = 10{\cdot}4$ cm.?

7. Find the radius of a sphere whose area is 52·41 sq. in.

8. Find the radius of a sphere whose volume is 112·8 cub. in.

9. The area of a triangle in terms of the lengths of its sides is given by

$$\sqrt{s(s-a)(s-b)(s-c)},$$

where s is half the sum of the sides $=\dfrac{a+b+c}{2}$.

Find the area of a triangle whose sides are 4·14 in., 3·762 in., 5·144 in.

10. Find the area of a triangle whose sides are 6·48 in., 4·76 in., 3·288 in.

TRIGONOMETRY

Trigonometrical Ratios.

This subject deals with another method of solving geometrical problems concerned with sides and angles of triangles.

In the section on relative velocity our answers were obtained by the method of drawing to scale and measuring lines and angles.

By the use of trigonometry the results can be more readily obtained and with far greater accuracy, since we eliminate all the errors due to protractor and ruler.

Suppose the line OP to be hinged at O and to revolve, with O as its centre, in an anticlockwise direction through 360° or one complete revolution. It will then have swept out a complete circle and returned to its original position.

While travelling from 0° to 90° it is said to be in the first quadrant.

From 90° to 180° it is in the second quadrant.

From 180° to 270° it is in the third quadrant.

From 270° to 360° it is in the fourth quadrant.

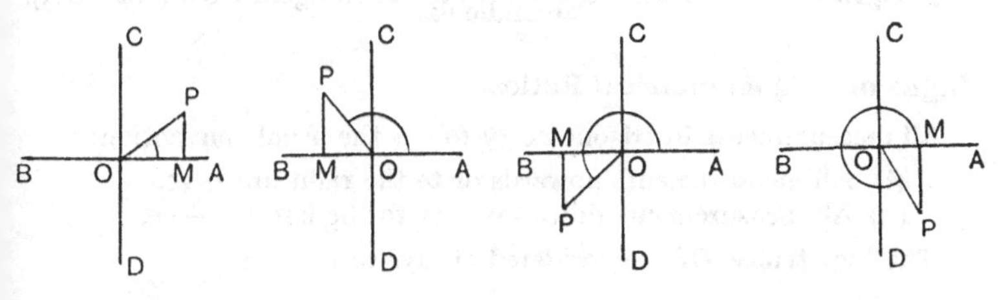

The angle POA, in trigonometry, is always given its full value, i.e. it is the complete angle through which the line OP has swept from its original position OA.

In the above four figures the line OP is shown stationary in each of the four quadrants in turn. In each case a perpendicular MP has been drawn to the horizontal AB. The full value of the angle POA in each quadrant is indicated.

If we consider any of the right-angled triangles OMP, so formed, we see that there are six separate ratios to each triangle with reference to the sides.

These ratios are $\frac{MP}{OP}, \frac{OP}{MP}, \frac{OM}{OP}, \frac{OP}{OM}, \frac{MP}{OM}, \frac{OM}{MP}$.

It will be seen that the second, fourth and sixth ratios are reciprocals of the first, third and fifth.

As in all right-angled triangles these sides have names: OP is the hypotenuse, MP is the perpendicular and OM is the base. And it will be seen at once that the ratio $\frac{MP}{OP}\left(\frac{\text{Perpendicular}}{\text{Hypotenuse}}\right)$ depends upon the size of the angle POA, i.e. the larger the angle POA the larger the ratio $\frac{MP}{OP}$. So that the six ratios mentioned above are all dependent on the size of the angle POA.

The six ratios are called the *trigonometrical* ratios of the angle POA and are different for different values of POA.

The ratio $\frac{MP}{OP}=\frac{\text{Perpendicular}}{\text{Hypotenuse}}$ is the sine of POA ($\sin POA$).

The ratio $\frac{OM}{OP}=\frac{\text{Base}}{\text{Hypotenuse}}$ is the cosine of POA ($\cos POA$).

The ratio $\frac{MP}{OM}=\frac{\text{Perpendicular}}{\text{Base}}$ is the tangent of POA ($\tan POA$).

And the reciprocals are:

Reciprocal of $\sin POA=\frac{\text{Hypotenuse}}{\text{Perpendicular}}=$ cosecant POA (cosec POA).

Reciprocal of $\cos POA=\frac{\text{Hypotenuse}}{\text{Base}}=$ secant POA (sec POA).

Reciprocal of $\tan POA=\frac{\text{Base}}{\text{Perpendicular}}=$ cotangent POA (cot POA).

Signs of Trigonometrical Ratios.

All measurements in trigonometry follow the usual convention:

(i) All measurements upwards or to the right are +ve.

(ii) All measurements downwards or to the left are −ve.

The hypotenuse OP is considered always to be +ve.

In measuring the base and perpendicular *always* start from O to M and then from M to P. In this way the co-ordinates are treated as "directed numbers" and the correct sign is automatically determined. For example, in the second quadrant

$$\cos POA \text{ is } \frac{OM \text{ (which is to the left or } -\text{ve)}}{OP \text{ (which is always } +\text{ve)}} = -\text{ve}.$$

In the third quadrant

$$\sin POA \text{ is } \frac{MP \text{ (which is downward or } -\text{ve)}}{OP \text{ (which is always } +\text{ve)}} = -\text{ve}.$$

Magnitude of Trigonometrical Ratios.

The magnitude of sin POA is seen to vary in the first quadrant from 0 to 1, as the length of the perpendicular MP grows from "nothing" to the length of OP.

It varies also in the second, third and fourth quadrants.

Check the following table for values and signs of the six trigonometrical ratios:

TABLE A

Ratio	Value of angle								
	0°	Sign	90°	Sign	180°	Sign	270°	Sign	360°
Sine	0	+	1	+	0	−	−1	−	0
Cosine	1	+	0	−	−1	−	0	+	1
Tangent	0	+	∞	−	0	+	∞	−	0
Cosecant	∞	+	1	+	∞	−	−1	−	∞
Secant	1	+	∞	−	−1	−	∞	+	1
Cotangent	∞	+	0	−	∞	+	0	−	∞

The symbol ∞ represents an infinitely large quantity, such as $\frac{OP}{\text{Nothing}}$.

Let us now evaluate the six trigonometrical ratios for some particular angle, say 36°.

Draw a line OA and make the angle $AOP = 36°$ (by protractor). Mark off $OP = 5$ cm. and draw MP perpendicular to AO. Measure OM and MP carefully and place the values on the lines.

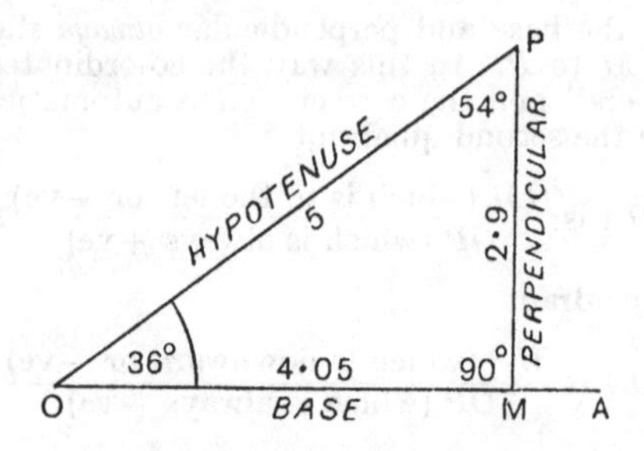

Write the names of the sides on the triangle, using $\angle POM$ (36°) as the angle of reference. Then from our measured values we get

$$\sin 36° = \frac{2{\cdot}9}{5} = 0{\cdot}58; \qquad \operatorname{cosec} 36° = \frac{5}{2{\cdot}9} = 1{\cdot}72;$$

$$\cos 36° = \frac{4{\cdot}05}{5} = 0{\cdot}81; \qquad \sec 36° = \frac{5}{4{\cdot}05} = 1{\cdot}23;$$

$$\tan 36° = \frac{2{\cdot}9}{4{\cdot}05} = 0{\cdot}716; \qquad \cot 36° = \frac{4{\cdot}05}{2{\cdot}9} = 1{\cdot}4.$$

These values are not to be compared in accuracy with those given in tables, but they afford good practice in protraction and measurement.

Tables and Logarithms of Trigonometrical Ratios.

At the end of the book, tables will be found showing:

(*a*) The natural values of all the six trigonometrical ratios for any angle from 0° to 90° at 1 min. intervals.

(*b*) The values of the logarithms of these ratios, also at 1 min. intervals.

These tables are similar in design to logarithm tables and have a "difference" column.

Suppose we wish to find the value of sin 20° 37′.

Turn to the tables headed SINES. Travel horizontally along the 20° line (on left) to column 30′. The figure tabulated is 0·3502. In the difference column the extra 7′ represent 19 to be added.

Hence sin 20° 37′ is 0·3521.

You will see that the SINES table is labelled COSINES at the bottom and degrees from 0° to 44° and 45° to 89° are marked on the *right* of each page, reading from bottom to top. The difference column is also at the bottom and reads from right to left. So that the tables will give values for cosines as well as sines if read backwards.

The tangent table will also give cotangents.

This is made possible because the sine of any angle A = cosine of the complement of A {i.e. $\cos(90° - A)$} and $\tan A = \cot(90° - A)$.

Cosecant and secant tables are omitted from this book because, if wanted, they can be obtained from sine and cosine tables, since

$$\operatorname{cosec} A = \frac{1}{\sin A} \quad \text{and} \quad \sec A = \frac{1}{\cos A}.$$

Be careful to see, especially with tangent and cotangent tables, whether the differences are added or subtracted. Also note that the signs + or − are not obtainable from the tables. They should be obtained from table A.

Check from the tables the values obtained for the trigonometrical ratios of 36°:

		By measurement	By tables
sin 36°	=	0·58	0·5878
cos 36°	=	0·81	0·8090
tan 36°	=	0·716	0·7265
cot 36°	=	1·4	1·3764

To obtain cosec 36° from the tables, we write

$$\operatorname{cosec} 36° = \frac{1}{\sin 36°}.$$

Taking logs, $\log \operatorname{cosec} 36° = \log 1 - \log \sin 36°$,

i.e. $\log \operatorname{cosec} 36° = 0 - \bar{1}{\cdot}7692$ (from log sine table)

$= 0{\cdot}2308.$

By antilogs, the answer is 1·702.

Similarly,

$$\sec 36° = \frac{1}{\cos 36°},$$

and from log cosine tables and antilog tables is found to be 1·236.

Thus:

		By measurement	By tables
cosec 36°	=	1·72	1·702
sec 36°	=	1·23	1·236

Exercise LX.

1. Give the complements of the following angles:

(*a*) 33° 10′. (*b*) 47° 25′. (*c*) 62° 47′. (*d*) 73° 21′.

2. Find the sine of each of the following angles and from the tables determine what angle has your answer as its cosine:

(*a*) 10° 20′. (*b*) 78° 40′. (*c*) 17° 23′.
(*d*) 32° 47′. (*e*) 31° 35′. (*f*) 67° 49′.

3. Find from the tables the cosines of the following:

(*a*) 14°. (*b*) 16° 20′. (*c*) 69° 50′.
(*d*) 32° 23′. (*e*) 38° 2′. (*f*) 49° 14′.

4. Give the log sine of (*a*) to (*d*) and the log cosine of (*e*) to (*h*):

(*a*) 22° 10′. (*b*) 28°. (*c*) 34° 15′. (*d*) 76° 21′.
(*e*) 64° 37′. (*f*) 26°. (*g*) 29° 20′. (*h*) 31° 3′.

5. Give the tangent of (*a*) to (*d*) and the cotangent of (*e*) to (*h*):

(*a*) 31°. (*b*) 42° 20′. (*c*) 57° 4′. (*d*) 68° 27′.
(*e*) 41°. (*f*) 43° 40′. (*g*) 52° 6′. (*h*) 39° 39′.

6. Give the log tan of (*a*) to (*d*) and the log cot of (*e*) to (*h*):

(*a*) 33°. (*b*) 42° 10′. (*c*) 51° 7′. (*d*) 35° 37′.
(*e*) 1°. (*f*) 5° 40′. (*g*) 16° 2′. (*h*) 37° 39′.

7. Using the log sine and log cosine tables and antilogs by log table, find:

(*a*) sin 31° 17′. (*b*) cos 25° 37′. (*c*) cosec 12°.
(*d*) sec 41°. (*e*) cosec 31° 3′. (*f*) sec 47° 6′.
(*g*) cosec 37° 37′. (*h*) sec 59° 57′.

Angles are said to be "supplementary" if when added together they equal 180°. Thus the supplement of 18° 31′ is

$$180° - 18° 31' = 161° 29'.$$

8. Give the supplements of these angles:

(*a*) 34°. (*b*) 42° 42′. (*c*) 91° 19′. (*d*) 57° 57′.

9. By drawing diagrams, state the sign of the following ratios:

(*a*) sin 120°. (*b*) tan 95°. (*c*) sec 102°. (*d*) cos 93°.
(*e*) tan 46°. (*f*) cosec 61°. (*g*) cosec 97°. (*h*) tan 120°.
(*i*) cotan 137°. (*j*) sec 34°.

Note (and illustrate by diagram) that

$\sin(180° - A) = \sin A$, $\cos(180° - A) = -\cos A$,
$\tan(180° - A) = -\tan A$, $\cot(180° - A) = -\cot A$,
$\sec(180° - A) = -\sec A$, $\operatorname{cosec}(180° - A) = \operatorname{cosec} A$.

10. Find the value of the following, including the sign:

(*a*) sin 120°. (*b*) cos 131° 10′. (*c*) tan 94° 18′.
(*d*) sec 128° 7′. (*e*) cot 112° 17′. (*f*) cot 16° 8′.
(*g*) cos 105° 57′. (*h*) cot 91° 49′. (*i*) sec 134° 17′.
(*j*) tan 161° 7′.

11. Write down the value only of:

(*a*) log sin 110° 6′. (*b*) log cos 20° 17′.
(*c*) log tan 67° 19′. (*d*) log sec 73° 17′.
(*e*) log cosec 131° 27′. (*f*) log cot 36° 17′.

12. What is the smallest angle whose

(*a*) sine is 0·3141; (*b*) cosine is 0·4261;
(*c*) tan is $-2{\cdot}163$; (*d*) cot is 2·819;
(*e*) cosec is 3·2222; (*f*) cosine is $-0{\cdot}3142$?

Terms used in Trigonometry.

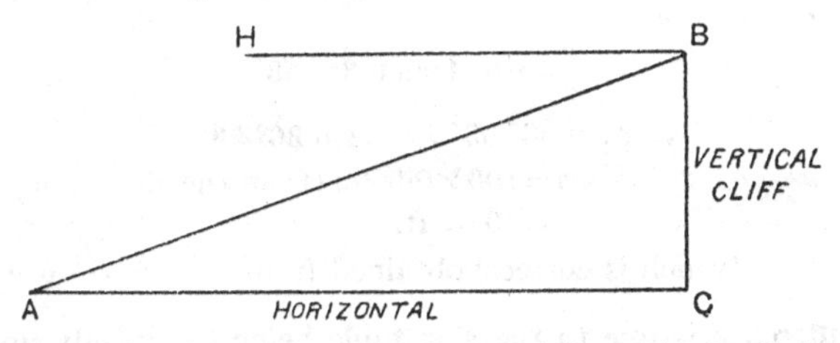

A vertical line from any point (C) coincides with the "plumb line" at that point.

A horizontal line at that point is a line drawn at right angles to the plumb line.

If an observer at A measures the angle between the horizontal and a cliff top B, then this angle BAC so measured is called the *Angle of Elevation* of the cliff top from A.

If the observer is on the cliff top at B and measures the angle between the horizontal BH and some object at A on the ground, then the angle ABH so measured is called the *Angle of Depression* from B to A. It will be seen that $\angle BAC = \angle ABH$.

Trigonometrical Solution of the Right-angled Triangle.

Let ABC be a triangle with a right angle at C. Then for purposes of brevity the angles may be referred to as angle A, angle B or angle C (using capital letters) and the sides opposite these angles by the corresponding small letters. Thus side AB is c, side AC is b, side BC is a.

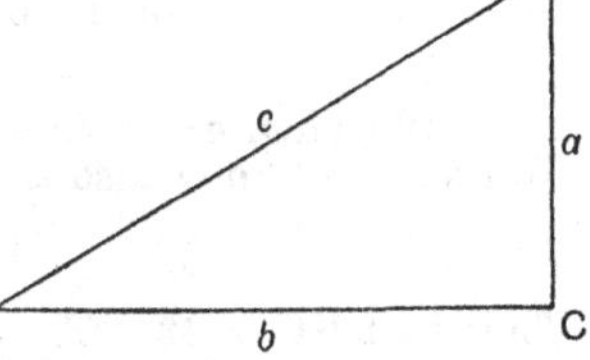

EXAMPLE 1. Suppose in the above triangle we are told that $C=90°$, $b=80$ ft. and $c=100$ ft. We are asked to find the angles A and B and the side a.

We can find a by the method of the "square of the hypotenuse".

$$c^2=b^2+a^2,$$

i.e.

$$a^2=c^2-b^2.$$

Therefore $a=\sqrt{c^2-b^2}=\sqrt{100^2-80^2}$

$$=\sqrt{10{,}000-6400}=\sqrt{3600}=60 \text{ ft.}$$

Let us now use a shorter method by trigonometry.

We know that $\cos A = \dfrac{\text{Base}}{\text{Hypotenuse}} = \dfrac{b}{c} = \dfrac{80}{100} = 0{\cdot}8.$

A reference to the cosine table shows that angle A is $36° 53'$.
Consequently B is the complement of A and is

$$90° - 36° 53' = 53° 7'.$$

We also know that

$$\frac{a}{c} = \sin A = \sin 36° 53'$$

and $$a = c \sin 36° 53' = 100 \sin 36° 53',$$

i.e. $$a = 100 \times 0{\cdot}6002 \text{ (from the sine table)}$$
$$= 60{\cdot}02 \text{ ft.}$$

The answer (which is correct) obtained from $a = \sqrt{c^2 - b^2}$ was exactly 60 ft.

The difference is due to the sine table being exact only up to four places of decimals.

The triangle is now solved, since we know A, B and C as well as a, b, c.

EXAMPLE 2. A triangle ABC, with right angle at C, has $A = 43° 30'$ and $c = 22{\cdot}4$ yd. Find the remaining sides and angle.

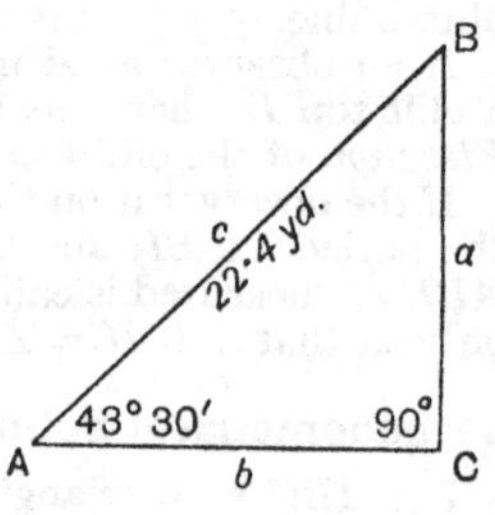

Draw the triangle and insert all known values, as above.

Then $B = 90° - 43° 30' = 46° 30'$.

And $\dfrac{a}{c} = \sin A$, i.e.

$$a = c \sin A = 22{\cdot}4 \times \sin 43° 30'.$$

Taking logs,

$$\log a = \log 22{\cdot}4 + \log \sin 43° 30'$$
$$= 1{\cdot}3502 + \bar{1}{\cdot}8378 = 1{\cdot}1880.$$

By antilogging, $a = 15{\cdot}42$ yd.
To find b, we know also that

$$\frac{b}{c} = \cos A.$$

Thus $b = 22{\cdot}4 \cos 43° 30'$.

Taking logs, $$\log b = \log 22{\cdot}4 + \log \cos 43° 30'$$
$$= 1{\cdot}3502 + \bar{1}{\cdot}8606 = 1{\cdot}2108.$$

By antilogging, $b = 16{\cdot}25$ yd.

EXAMPLE 3. An observer in a boat approaching the coast observes the angle of elevation to a cliff top to be $5° 30'$ by sextant. The chart shows the cliff to be 300 ft. high. How far is the boat to seaward of the cliff?

Draw a diagram and place dimensions on it.
The side of triangle b is the distance we wish to find.

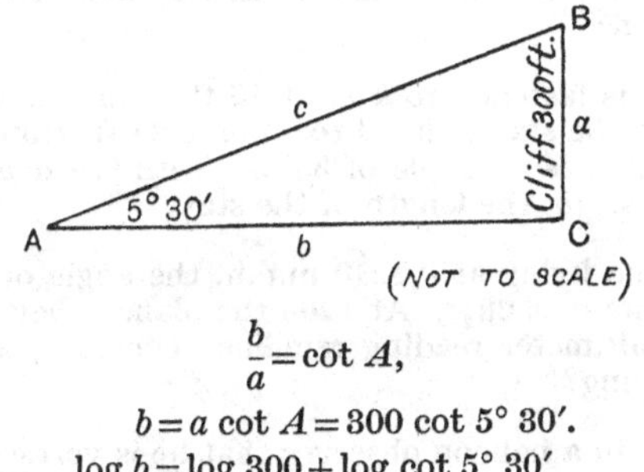

Therefore $\frac{b}{a} = \cot A,$

$$b = a \cot A = 300 \cot 5° \, 30'.$$

By logs, $\log b = \log 300 + \log \cot 5° \, 30'$
$= 2{\cdot}4771 + 1{\cdot}0164$
$= 3{\cdot}4935.$

By antilogs, $b = 3115$ ft.
Distance of boat from shore = 3115 ft.

EXERCISE LXI.

(Draw diagrams to help solve all these questions.)

1. Solve the following right-angled triangles if C is the right angle:

	Parts given	Parts required
(a)	$a = 33{\cdot}96,\ c = 37{\cdot}4$	A, B, b
(b)	$A = 34° \, 15',\ a = 421{\cdot}6$	B, c, b
(c)	$B = 47° \, 15',\ c = 4{\cdot}614$	A, a, b
(d)	$a = 10,\ b = 4$	A, B, c
(e)	$B = 68° \, 50',\ a = 364{\cdot}6$	A, b, c
(f)	$b = 1{\cdot}437,\ c = 3{\cdot}465$	A, B, a
(g)	$A = 43° \, 30',\ c = 22{\cdot}4$	B, a, b
(h)	$a = 49{\cdot}34,\ b = 66{\cdot}04$	A, B, c
(i)	$a = 3555,\ b = 2354$	A, B, c
(j)	$c = 45,\ b = 35{\cdot}76$	$A, B, a.$

2. A boat is $1\frac{1}{2}$ cables to seaward of a cliff whose top is charted as 347 ft. above sea level. What sextant angle will the cliff top subtend to an observer in the boat (1 cable = 200 yd.)?

3. The angle of elevation of the sun on a certain day was 50° 13′. How long a shadow did a vertical pole 11 ft. in length cast on the horizontal ground?

4. At 1210 hours a plane flies from a certain position on a course 063° T with G/S 73 m.p.h. How many miles (a) North, (b) East of the departure point is the plane at 1340 hours?

5. A pilot flying at G/S 90 m.p.h. on a course 095° T observes a church to bear 005° T at time 1000. At 1010 the same church was bearing Red 123°. How far was the plane from the church when the second bearing was taken?

6. A wire stay is fastened to a mast 15 ft. from the top of the mast. The other end of the stay is fixed to a stake 23 ft. from the foot of the mast. The stay makes an angle of 33° 30′ with the mast. Find (*a*) the height of the mast, (*b*) the length of the stay.

7. From a plane flying at G/S 80 m.p.h. the angle of depression of a church at 1200 hours is $23\frac{1}{2}°$. At 1206 the plane is vertically above the church. If the altimeter reading remained constant, at what altitude was the plane flying?

8. An observer in a balloon observes that he is vertically over a point midway between two ground objects 2000 yd. apart. These two objects subtend an angle of 110° from his position. At what height is the balloon?

9. A pilot flying at a low altitude on a course 000 T with G/S 60 m.p.h. observes a tower to bear 030° T at 1100 hours. At 1230 the bearing of the same tower is 120° T. Find (*a*) the plane's nearest approach to the tower, (*b*) the time when the plane was nearest to the tower.

10. A pilot leaves an aerodrome at 2200 hours, and flies with G/S 90 m.p.h. on a course 031° T. At 2240 he A/CO to 121° T and reduces speed to G/S 80 m.p.h. If at 2310 he sets a course straight for home, what true course must he steer? Approximate your answer to the nearest half degree.

11. An aircraft flies 261 nautical miles due E. from lat. 49° 24′ N. Estimate the change in longitude of arrival point from the formula:

Change of long. (in min.) = Easting in N.M. × secant of lat.

12. A radio station X is in position 52° 30′ N., 4° 20′ E. and an aircraft is in position 47° 30′ N., 8° 40′ W. Determine the magnitude of the convergency from this equation:

Convergency in degrees = change of long. in degrees
× sine of mean latitude.

13. At 1420 hr. an aircraft sets out on a track of 064° T with ground speed 147 m.p.h. How many miles east of the departure point is the aircraft at 1512 hr.?

14. Find, without the use of logarithms, the value of

$$\frac{1-\cos^2 42^\circ\, 11'}{(1-\cos 42^\circ\, 11')^2}.$$

15. A pilot flying on a straight course at a low altitude observes a radio mast to bear Green 029° at 1458 hr. At 1501 hr. the bearing of the same mast is Green 058°. If the ground speed of the aircraft is 220 m.p.h. what is its nearest distance of approach to the mast? (Answer in miles to three decimal places.)

16. In astro-navigation use is made of haversines, where

$$\text{Haversine } A = \tfrac{1}{2}(1 - \cos A).$$

What is the value of $\sqrt{\text{haversine } 21^\circ\ 20'}$?

17. From the bridge of a ship, 30 ft. above sea level, the angle of elevation to a light on a vertical cliff top is 12° 30′. The light is charted as 455 ft. above sea level. How far off shore is the ship? (Answer to nearest yard.)

18. A seaplane mooring buoy is to be anchored at one cable length to seaward of a vertical cliff 131 ft. high. What sextant angle will the buoy subtend from the nearest point on the cliff top when in correct position?

19. On a certain day at noon the length of the shadow of a ship's mast on the deck was 41 ft. 3 in. If the height of the mast was 52 ft. 6 in. what was the sun's altitude above the horizon?

20. An observer in an aircraft flying at an altitude of 8500 ft. reads the angle of depression to an aerodrome in the direct line of flight as 23° 20′. In 1 min. 6 sec. the aircraft is vertically above the aerodrome. What is the speed of the aircraft in m.p.h.?

											Differences.								
	0	1	2	3	4	5	6	7	8	9	1	2	3	4	5	6	7	8	9
10	0000	0043	0086	0128	0170	0212	0253	0294	0334	0374	4	8	12	17	21	25	29	33	37
11	0414	0453	0492	0531	0569	0607	0645	0682	0719	0755	4	8	11	15	19	23	26	30	34
12	0792	0828	0864	0899	0934	0969	1004	1038	1072	1106	3	7	10	14	17	21	24	28	31
13	1139	1173	1206	1239	1271	1303	1335	1367	1399	1430	3	6	10	13	16	19	23	26	29
14	1461	1492	1523	1553	1584	1614	1644	1673	1703	1732	3	6	9	12	15	18	21	24	27
15	1761	1790	1818	1847	1875	1903	1931	1959	1987	2014	3	6	8	11	14	17	20	22	25
16	2041	2068	2095	2122	2148	2175	2201	2227	2253	2279	3	5	8	11	13	16	18	21	24
17	2304	2330	2355	2380	2405	2430	2455	2480	2504	2529	2	5	7	10	12	15	17	20	22
18	2553	2577	2601	2625	2648	2672	2695	2718	2742	2765	2	5	7	9	12	14	16	19	21
19	2788	2810	2833	2856	2878	2900	2923	2945	2967	2989	2	4	7	9	11	13	16	18	20
20	3010	3032	3054	3075	3096	3118	3139	3160	3181	3201	2	4	6	8	11	13	15	17	19
21	3222	3243	3263	3284	3304	3324	3345	3365	3385	3404	2	4	6	8	10	12	14	16	18
22	3424	3444	3464	3483	3502	3522	3541	3560	3579	3598	2	4	6	8	10	12	14	15	17
23	3617	3636	3655	3674	3692	3711	3729	3747	3766	3784	2	4	6	7	9	11	13	15	17
24	3802	3820	3838	3856	3874	3892	3909	3927	3945	3962	2	4	5	7	9	11	12	14	16
25	3979	3997	4014	4031	4048	4065	4082	4099	4116	4133	2	3	5	7	9	10	12	14	15
26	4150	4166	4183	4200	4216	4232	4249	4265	4281	4298	2	3	5	7	8	10	11	13	15
27	4314	4330	4346	4362	4378	4393	4409	4425	4440	4456	2	3	5	6	8	9	11	13	14
28	4472	4487	4502	4518	4533	4548	4564	4579	4594	4609	2	3	5	6	8	9	11	12	14
29	4624	4639	4654	4669	4683	4698	4713	4728	4742	4757	1	3	4	6	7	9	10	12	13
30	4771	4786	4800	4814	4829	4843	4857	4871	4886	4900	1	3	4	6	7	9	10	11	13
31	4914	4928	4942	4955	4969	4983	4997	5011	5024	5038	1	3	4	6	7	8	10	11	12
32	5051	5065	5079	5092	5105	5119	5132	5145	5159	5172	1	3	4	5	7	8	9	11	12
33	5185	5198	5211	5224	5237	5250	5263	5276	5289	5302	1	3	4	5	6	8	9	10	12
34	5315	5328	5340	5353	5366	5378	5391	5403	5416	5428	1	3	4	5	6	8	9	10	11
35	5441	5453	5465	5478	5490	5502	5514	5527	5539	5551	1	2	4	5	6	7	9	10	11
36	5563	5575	5587	5599	5611	5623	5635	5647	5658	5670	1	2	4	5	6	7	8	10	11
37	5682	5694	5705	5717	5729	5740	5752	5763	5775	5786	1	2	3	5	6	7	8	9	10
38	5798	5809	5821	5832	5843	5855	5866	5877	5888	5899	1	2	3	5	6	7	8	9	10
39	5911	5922	5933	5944	5955	5966	5977	5988	5999	6010	1	2	3	4	5	7	8	9	10
40	6021	6031	6042	6053	6064	6075	6085	6096	6107	6117	1	2	3	4	5	6	8	9	10
41	6128	6138	6149	6160	6170	6180	6191	6201	6212	6222	1	2	3	4	5	6	7	8	9
42	6232	6243	6253	6263	6274	6284	6294	6304	6314	6325	1	2	3	4	5	6	7	8	9
43	6335	6345	6355	6365	6375	6385	6395	6405	6415	6425	1	2	3	4	5	6	7	8	9
44	6435	6444	6454	6464	6474	6484	6493	6503	6513	6522	1	2	3	4	5	6	7	8	9
45	6532	6542	6551	6561	6571	6580	6590	6599	6609	6618	1	2	3	4	5	6	7	8	9
46	6628	6637	6646	6656	6665	6675	6684	6693	6702	6712	1	2	3	4	5	6	7	7	8
47	6721	6730	6739	6749	6758	6767	6776	6785	6794	6803	1	2	3	4	5	5	6	7	8
48	6812	6821	6830	6839	6848	6857	6866	6875	6884	6893	1	2	3	4	4	5	6	7	8
49	6902	6911	6920	6928	6937	6946	6955	6964	6972	6981	1	2	3	4	4	5	6	7	8
50	6990	6998	7007	7016	7024	7033	7042	7050	7059	7067	1	2	3	3	4	5	6	7	8
51	7076	7084	7093	7101	7110	7118	7126	7135	7143	7152	1	2	3	3	4	5	6	7	8
52	7160	7168	7177	7185	7193	7202	7210	7218	7226	7235	1	2	2	3	4	5	6	7	7
53	7243	7251	7259	7267	7275	7284	7292	7300	7308	7316	1	2	2	3	4	5	6	6	7
54	7324	7332	7340	7348	7356	7364	7372	7380	7388	7396	1	2	2	3	4	5	6	6	7
	0	1	2	3	4	5	6	7	8	9	1	2	3	4	5	6	7	8	9

	0	1	2	3	4	5	6	7	8	9	Differences. 1	2	3	4	5	6	7	8	9
55	7404	7412	7419	7427	7435	7443	7451	7459	7466	7474	1	2	2	3	4	5	5	6	7
56	7482	7490	7497	7505	7513	7520	7528	7536	7543	7551	1	2	2	3	4	5	5	6	7
57	7559	7566	7574	7582	7589	7597	7604	7612	7619	7627	1	2	2	3	4	5	5	6	7
58	7634	7642	7649	7657	7664	7672	7679	7686	7694	7701	1	1	2	3	4	4	5	6	7
59	7709	7716	7723	7731	7738	7745	7752	7760	7767	7774	1	1	2	3	4	4	5	6	7
60	7782	7789	7796	7803	7810	7818	7825	7832	7839	7846	1	1	2	3	4	4	5	6	6
61	7853	7860	7868	7875	7882	7889	7896	7903	7910	7917	1	1	2	3	4	4	5	6	6
62	7924	7931	7938	7945	7952	7959	7966	7973	7980	7987	1	1	2	3	3	4	5	6	6
63	7993	8000	8007	8014	8021	8028	8035	8041	8048	8055	1	1	2	3	3	4	5	5	6
64	8062	8069	8075	8082	8089	8096	8102	8109	8116	8122	1	1	2	3	3	4	5	5	6
65	8129	8136	8142	8149	8156	8162	8169	8176	8182	8189	1	1	2	3	3	4	5	5	6
66	8195	8202	8209	8215	8222	8228	8235	8241	8248	8254	1	1	2	3	3	4	5	5	6
67	8261	8267	8274	8280	8287	8293	8299	8306	8312	8319	1	1	2	3	3	4	5	5	6
68	8325	8331	8338	8344	8351	8357	8363	8370	8376	8382	1	1	2	3	3	4	4	5	6
69	8388	8395	8401	8407	8414	8420	8426	8432	8439	8445	1	1	2	2	3	4	4	5	6
70	8451	8457	8463	8470	8476	8482	8488	8494	8500	8506	1	1	2	2	3	4	4	5	6
71	8513	8519	8525	8531	8537	8543	8549	8555	8561	8567	1	1	2	2	3	4	4	5	5
72	8573	8579	8585	8591	8597	8603	8609	8615	8621	8627	1	1	2	2	3	4	4	5	5
73	8633	8639	8645	8651	8657	8663	8669	8675	8681	8686	1	1	2	2	3	4	4	5	5
74	8692	8698	8704	8710	8716	8722	8727	8733	8739	8745	1	1	2	2	3	4	4	5	5
75	8751	8756	8762	8768	8774	8779	8785	8791	8797	8802	1	1	2	2	3	3	4	5	5
76	8808	8814	8820	8825	8831	8837	8842	8848	8854	8859	1	1	2	2	3	3	4	5	5
77	8865	8871	8876	8882	8887	8893	8899	8904	8910	8915	1	1	2	2	3	3	4	4	5
78	8921	8927	8932	8938	8943	8949	8954	8960	8965	8971	1	1	2	2	3	3	4	4	5
79	8976	8982	8987	8993	8998	9004	9009	9015	9020	9025	1	1	2	2	3	3	4	4	5
80	9031	9036	9042	9047	9053	9058	9063	9069	9074	9079	1	1	2	2	3	3	4	4	5
81	9085	9090	9096	9101	9106	9112	9117	9122	9128	9133	1	1	2	2	3	3	4	4	5
82	9138	9143	9149	9154	9159	9165	9170	9175	9180	9186	1	1	2	2	3	3	4	4	5
83	9191	9196	9201	9206	9212	9217	9222	9227	9232	9238	1	1	2	2	3	3	4	4	5
84	9243	9248	9253	9258	9263	9269	9274	9279	9284	9289	1	1	2	2	3	3	4	4	5
85	9294	9299	9304	9309	9315	9320	9325	9330	9335	9340	1	1	2	2	3	3	4	4	5
86	9345	9350	9355	9360	9365	9370	9375	9380	9385	9390	1	1	1	2	3	3	4	4	5
87	9395	9400	9405	9410	9415	9420	9425	9430	9435	9440	0	1	1	2	2	3	3	4	4
88	9445	9450	9455	9460	9465	9469	9474	9479	9484	9489	0	1	1	2	2	3	3	4	4
89	9494	9499	9504	9509	9513	9518	9523	9528	9533	9538	0	1	1	2	2	3	3	4	4
90	9542	9547	9552	9557	9562	9566	9571	9576	9581	9586	0	1	1	2	2	3	3	4	4
91	9590	9595	9600	9605	9609	9614	9619	9624	9628	9633	0	1	1	2	2	3	3	4	4
92	9638	9643	9647	9652	9657	9661	9666	9671	9675	9680	0	1	1	2	2	3	3	4	4
93	9685	9689	9694	9699	9703	9708	9713	9717	9722	9727	0	1	1	2	2	3	3	4	4
94	9731	9736	9741	9745	9750	9754	9759	9763	9768	9773	0	1	1	2	2	3	3	4	4
95	9777	9782	9786	9791	9795	9800	9805	9809	9814	9818	0	1	1	2	2	3	3	4	4
96	9823	9827	9832	9836	9841	9845	9850	9854	9859	9863	0	1	1	2	2	3	3	4	4
97	9868	9872	9877	9881	9886	9890	9894	9899	9903	9908	0	1	1	2	2	3	3	4	4
98	9912	9917	9921	9926	9930	9934	9939	9943	9948	9952	0	1	1	2	2	3	3	4	4
99	9956	9961	9965	9969	9974	9978	9983	9987	9991	9996	0	1	1	2	2	3	3	3	4
	0	1	2	3	4	5	6	7	8	9	1	2	3	4	5	6	7	8	9

	0	1	2	3	4	5	6	7	8	9	Differences								
											1	2	3	4	5	6	7	8	9
·00	1000	1002	1005	1007	1009	1012	1014	1016	1019	1021	0	0	1	1	1	1	2	2	2
·01	1023	1026	1028	1030	1033	1035	1038	1040	1042	1045	0	0	1	1	1	1	2	2	2
·02	1047	1050	1052	1054	1057	1059	1062	1064	1067	1069	0	0	1	1	1	1	2	2	2
·03	1072	1074	1076	1079	1081	1084	1086	1089	1091	1094	0	0	1	1	1	1	2	2	2
·04	1096	1099	1102	1104	1107	1109	1112	1114	1117	1119	0	1	1	1	1	2	2	2	2
·05	1122	1125	1127	1130	1132	1135	1138	1140	1143	1146	0	1	1	1	1	2	2	2	2
·06	1148	1151	1153	1156	1159	1161	1164	1167	1169	1172	0	1	1	1	1	2	2	2	2
·07	1175	1178	1180	1183	1186	1189	1191	1194	1197	1199	0	1	1	1	1	2	2	2	2
·08	1202	1205	1208	1211	1213	1216	1219	1222	1225	1227	0	1	1	1	1	2	2	2	3
·09	1230	1233	1236	1239	1242	1245	1247	1250	1253	1256	0	1	1	1	1	2	2	2	3
·10	1259	1262	1265	1268	1271	1274	1276	1279	1282	1285	0	1	1	1	1	2	2	2	3
·11	1288	1291	1294	1297	1300	1303	1306	1309	1312	1315	0	1	1	1	2	2	2	2	3
·12	1318	1321	1324	1327	1330	1334	1337	1340	1343	1346	0	1	1	1	2	2	2	2	3
·13	1349	1352	1355	1358	1361	1365	1368	1371	1374	1377	0	1	1	1	2	2	2	3	3
·14	1380	1384	1387	1390	1393	1396	1400	1403	1406	1409	0	1	1	1	2	2	2	3	3
·15	1413	1416	1419	1422	1426	1429	1432	1435	1439	1442	0	1	1	1	2	2	2	3	3
·16	1445	1449	1452	1455	1459	1462	1466	1469	1472	1476	0	1	1	1	2	2	2	3	3
·17	1479	1483	1486	1489	1493	1496	1500	1503	1507	1510	0	1	1	1	2	2	2	3	3
·18	1514	1517	1521	1524	1528	1531	1535	1538	1542	1545	0	1	1	1	2	2	2	3	3
·19	1549	1552	1556	1560	1563	1567	1570	1574	1578	1581	0	1	1	1	2	2	3	3	3
·20	1585	1589	1592	1596	1600	1603	1607	1611	1614	1618	0	1	1	1	2	2	3	3	3
·21	1622	1626	1629	1633	1637	1641	1644	1648	1652	1656	0	1	1	2	2	2	3	3	3
·22	1660	1663	1667	1671	1675	1679	1683	1687	1690	1694	0	1	1	2	2	2	3	3	3
·23	1698	1702	1706	1710	1714	1718	1722	1726	1730	1734	0	1	1	2	2	2	3	3	4
·24	1738	1742	1746	1750	1754	1758	1762	1766	1770	1774	0	1	1	2	2	2	3	3	4
·25	1778	1782	1786	1791	1795	1799	1803	1807	1811	1816	0	1	1	2	2	2	3	3	4
·26	1820	1824	1828	1832	1837	1841	1845	1849	1854	1858	0	1	1	2	2	3	3	3	4
·27	1862	1866	1871	1875	1879	1884	1888	1892	1897	1901	0	1	1	2	2	3	3	3	4
·28	1905	1910	1914	1919	1923	1928	1932	1936	1941	1945	0	1	1	2	2	3	3	4	4
·29	1950	1954	1959	1963	1968	1972	1977	1982	1986	1991	0	1	1	2	2	3	3	4	4
·30	1995	2000	2004	2009	2014	2018	2023	2028	2032	2037	0	1	1	2	2	3	3	4	4
·31	2042	2046	2051	2056	2061	2065	2070	2075	2080	2084	0	1	1	2	2	3	3	4	4
·32	2089	2094	2099	2104	2109	2113	2118	2123	2128	2133	0	1	1	2	2	3	3	4	4
·33	2138	2143	2148	2153	2158	2163	2168	2173	2178	2183	0	1	1	2	2	3	3	4	4
·34	2188	2193	2198	2203	2208	2213	2218	2223	2228	2234	1	1	2	2	3	3	4	4	5
·35	2239	2244	2249	2254	2259	2265	2270	2275	2280	2286	1	1	2	2	3	3	4	4	5
·36	2291	2296	2301	2307	2312	2317	2323	2328	2333	2339	1	1	2	2	3	3	4	4	5
·37	2344	2350	2355	2360	2366	2371	2377	2382	2388	2393	1	1	2	2	3	3	4	4	5
·38	2399	2404	2410	2415	2421	2427	2432	2438	2443	2449	1	1	2	2	3	3	4	4	5
·39	2455	2460	2466	2472	2477	2483	2489	2495	2500	2506	1	1	2	2	3	3	4	5	5
·40	2512	2518	2523	2529	2535	2541	2547	2553	2559	2564	1	1	2	2	3	4	4	5	5
·41	2570	2576	2582	2588	2594	2600	2606	2612	2618	2624	1	1	2	2	3	4	4	5	5
·42	2630	2636	2642	2649	2655	2661	2667	2673	2679	2685	1	1	2	2	3	4	4	5	6
·43	2692	2698	2704	2710	2716	2723	2729	2735	2742	2748	1	1	2	3	3	4	4	5	6
·44	2754	2761	2767	2773	2780	2786	2793	2799	2805	2812	1	1	2	3	3	4	4	5	6
·45	2818	2825	2831	2838	2844	2851	2858	2864	2871	2877	1	1	2	3	3	4	5	5	6
·46	2884	2891	2897	2904	2911	2917	2924	2931	2938	2944	1	1	2	3	3	4	5	5	6
·47	2951	2958	2965	2972	2979	2985	2992	2999	3006	3013	1	1	2	3	3	4	5	5	6
·48	3020	3027	3034	3041	3048	3055	3062	3069	3076	3083	1	1	2	3	4	4	5	6	6
·49	3090	3097	3105	3112	3119	3126	3133	3141	3148	3155	1	1	2	3	4	4	5	6	6

	0	1	2	3	4	5	6	7	8	9	Differences								
											1	2	3	4	5	6	7	8	9
·50	3162	3170	3177	3184	3192	3199	3206	3214	3221	3228	1	1	2	3	4	4	5	6	7
·51	3236	3243	3251	3258	3266	3273	3281	3289	3296	3304	1	2	2	3	4	5	5	6	7
·52	3311	3319	3327	3334	3342	3350	3357	3365	3373	3381	1	2	2	3	4	5	5	6	7
·53	3388	3396	3404	3412	3420	3428	3436	3443	3451	3459	1	2	2	3	4	5	6	6	7
·54	3467	3475	3483	3491	3499	3508	3516	3524	3532	3540	1	2	2	3	4	5	6	6	7
·55	3548	3556	3565	3573	3581	3589	3597	3606	3614	3622	1	2	2	3	4	5	6	7	7
·56	3631	3639	3648	3656	3664	3673	3681	3690	3698	3707	1	2	3	3	4	5	6	7	8
·57	3715	3724	3733	3741	3750	3758	3767	3776	3784	3793	1	2	3	3	4	5	6	7	8
·58	3802	3811	3819	3828	3837	3846	3855	3864	3873	3882	1	2	3	4	4	5	6	7	8
·59	3890	3899	3908	3917	3926	3936	3945	3954	3963	3972	1	2	3	4	5	5	6	7	8
·60	3981	3990	3999	4009	4018	4027	4036	4046	4055	4064	1	2	3	4	5	6	6	7	8
·61	4074	4083	4093	4102	4111	4121	4130	4140	4150	4159	1	2	3	4	5	6	7	8	9
·62	4169	4178	4188	4198	4207	4217	4227	4236	4246	4256	1	2	3	4	5	6	7	8	9
·63	4266	4276	4285	4295	4305	4315	4325	4335	4345	4355	1	2	3	4	5	6	7	8	9
·64	4365	4375	4385	4395	4406	4416	4426	4436	4446	4457	1	2	3	4	5	6	7	8	9
·65	4467	4477	4487	4498	4508	4519	4529	4539	4550	4560	1	2	3	4	5	6	7	8	9
·66	4571	4581	4592	4603	4613	4624	4634	4645	4656	4667	1	2	3	4	5	6	7	8	10
·67	4677	4688	4699	4710	4721	4732	4742	4753	4764	4775	1	2	3	4	5	7	8	9	10
·68	4786	4797	4808	4819	4831	4842	4853	4864	4875	4887	1	2	3	4	6	7	8	9	10
·69	4898	4909	4920	4932	4943	4955	4966	4977	4989	5000	1	2	3	5	6	7	8	9	10
·70	5012	5023	5035	5047	5058	5070	5082	5093	5105	5117	1	2	4	5	6	7	8	9	11
·71	5129	5140	5152	5164	5176	5188	5200	5212	5224	5236	1	2	4	5	6	7	8	10	11
·72	5248	5260	5272	5284	5297	5309	5321	5333	5346	5358	1	2	4	5	6	7	9	10	11
·73	5370	5383	5395	5408	5420	5433	5445	5458	5470	5483	1	3	4	5	6	8	9	10	11
·74	5495	5508	5521	5534	5546	5559	5572	5585	5598	5610	1	3	4	5	6	8	9	10	12
·75	5623	5636	5649	5662	5675	5689	5702	5715	5728	5741	1	3	4	5	7	8	9	10	12
·76	5754	5768	5781	5794	5808	5821	5834	5848	5861	5875	1	3	4	5	7	8	9	11	12
·77	5888	5902	5916	5929	5943	5957	5970	5984	5998	6012	1	3	4	5	7	8	10	11	12
·78	6026	6039	6053	6067	6081	6095	6109	6124	6138	6152	1	3	4	6	7	8	10	11	13
·79	6166	6180	6194	6209	6223	6237	6252	6266	6281	6295	1	3	4	6	7	9	10	11	13
·80	6310	6324	6339	6353	6368	6383	6397	6412	6427	6442	1	3	4	6	7	9	10	12	13
·81	6457	6471	6486	6501	6516	6531	6546	6561	6577	6592	2	3	5	6	8	9	11	12	14
·82	6607	6622	6637	6653	6668	6683	6699	6714	6730	6745	2	3	5	6	8	9	11	12	14
·83	6761	6776	6792	6808	6823	6839	6855	6871	6887	6902	2	3	5	6	8	9	11	13	14
·84	6918	6934	6950	6966	6982	6998	7015	7031	7047	7063	2	3	5	6	8	10	11	13	15
·85	7079	7096	7112	7129	7145	7161	7178	7194	7211	7228	2	3	5	7	8	10	12	13	15
·86	7244	7261	7278	7295	7311	7328	7345	7362	7379	7396	2	3	5	7	8	10	12	13	15
·87	7413	7430	7447	7464	7482	7499	7516	7534	7551	7568	2	3	5	7	9	10	12	14	16
·88	7586	7603	7621	7638	7656	7674	7691	7709	7727	7745	2	4	5	7	9	11	12	14	16
·89	7762	7780	7798	7816	7834	7852	7870	7889	7907	7925	2	4	5	7	9	11	13	14	16
·90	7943	7962	7980	7998	8017	8035	8054	8072	8091	8110	2	4	6	7	9	11	13	15	17
·91	8128	8147	8166	8185	8204	8222	8241	8260	8279	8299	2	4	6	8	9	11	13	15	17
·92	8318	8337	8356	8375	8395	8414	8433	8453	8472	8492	2	4	6	8	10	12	14	15	17
·93	8511	8531	8551	8570	8590	8610	8630	8650	8670	8690	2	4	6	8	10	12	14	16	18
·94	8710	8730	8750	8770	8790	8810	8831	8851	8872	8892	2	4	6	8	10	12	14	16	18
·95	8913	8933	8954	8974	8995	9016	9036	9057	9078	9099	2	4	6	8	10	12	14	17	19
·96	9120	9141	9162	9183	9204	9226	9247	9268	9290	9311	2	4	6	8	11	13	15	17	19
·97	9333	9354	9376	9397	9419	9441	9462	9484	9506	9528	2	4	7	9	11	13	15	17	20
·98	9550	9572	9594	9616	9638	9661	9683	9705	9727	9750	2	4	7	9	11	13	16	18	20
·99	9772	9795	9817	9840	9863	9886	9908	9931	9954	9977	2	5	7	9	11	14	16	18	20

SINES.

	0′	10′	20′	30′	40′	50′	60′		1′	2′	3′	4′	5′	6′	7′	8′	9′
									Differences.								
0°	0·0000	0·0029	0·0058	0·0087	0·0116	0·0145	0·0175	89°	3	6	9	12	15	17	20	23	26
1	·0175	·0204	·0233	·0262	·0291	·0320	·0349	88	3	6	9	12	15	17	20	23	26
2	·0349	·0378	·0407	·0436	·0465	·0494	·0523	87	3	6	9	12	15	17	20	23	26
3	·0523	·0552	·0581	·0610	·0640	·0669	·0698	86	3	6	9	12	15	17	20	23	26
4	·0698	·0727	·0756	·0785	·0814	·0843	·0872	85	3	6	9	12	15	17	20	23	26
5	0·0872	0·0901	0·0929	0·0958	0·0987	0·1016	0·1045	84	3	6	9	12	14	17	20	23	26
6	·1045	·1074	·1103	·1132	·1161	·1190	·1219	83	3	6	9	12	14	17	20	23	26
7	·1219	·1248	·1276	·1305	·1334	·1363	·1392	82	3	6	9	12	14	17	20	23	26
8	·1392	·1421	·1449	·1478	·1507	·1536	·1564	81	3	6	9	11	14	17	20	23	26
9	·1564	·1593	·1622	·1650	·1679	·1708	·1736	80°	3	6	9	11	14	17	20	23	26
10°	0·1736	0·1765	0·1794	0·1822	0·1851	0·1880	0·1908	79	3	6	9	11	14	17	20	23	26
11	·1908	·1937	·1965	·1994	·2022	·2051	·2079	78	3	6	9	11	14	17	20	23	26
12	·2079	·2108	·2136	·2164	·2193	·2221	·2250	77	3	6	9	11	14	17	20	23	26
13	·2250	·2278	·2306	·2334	·2363	·2391	·2419	76	3	6	8	11	14	17	20	23	25
14	·2419	·2447	·2476	·2504	·2532	·2560	·2588	75	3	6	8	11	14	17	20	23	25
15	0·2588	0·2616	0·2644	0·2672	0·2700	0·2728	0·2756	74	3	6	8	11	14	17	20	22	25
16	·2756	·2784	·2812	·2840	·2868	·2896	·2924	73	3	6	8	11	14	17	20	22	25
17	·2924	·2952	·2979	·3007	·3035	·3062	·3090	72	3	6	8	11	14	17	19	22	25
18	·3090	·3118	·3145	·3173	·3201	·3228	·3256	71	3	6	8	11	14	17	19	22	25
19	·3256	·3283	·3311	·3338	·3365	·3393	·3420	70°	3	5	8	11	14	16	19	22	25
20°	0·3420	0·3448	0·3475	0·3502	0·3529	0·3557	0·3584	69	3	5	8	11	14	16	19	22	25
21	·3584	·3611	·3638	·3665	·3692	·3719	·3746	68	3	5	8	11	14	16	19	22	24
22	·3746	·3773	·3800	·3827	·3854	·3881	·3907	67	3	5	8	11	13	16	19	21	24
23	·3907	·3934	·3961	·3987	·4014	·4041	·4067	66	3	5	8	11	13	16	19	21	24
24	·4067	·4094	·4120	·4147	·4173	·4200	·4226	65	3	5	8	11	13	16	19	21	24
25	0·4226	0·4253	0·4279	0·4305	0·4331	0·4358	0·4384	64	3	5	8	10	13	16	18	21	24
26	·4384	·4410	·4436	·4462	·4488	·4514	·4540	63	3	5	8	10	13	16	18	21	23
27	·4540	·4566	·4592	·4617	·4643	·4669	·4695	62	3	5	8	10	13	15	18	21	23
28	·4695	·4720	·4746	·4772	·4797	·4823	·4848	61	3	5	8	10	13	15	18	20	23
29	·4848	·4874	·4899	·4924	·4950	·4975	·5000	60°	3	5	8	10	13	15	18	20	23
30°	0·5000	0·5025	0·5050	0·5075	0·5100	0·5125	0·5150	59	3	5	8	10	13	15	18	20	23
31	·5150	·5175	·5200	·5225	·5250	·5275	·5299	58	2	5	7	10	12	15	17	20	22
32	·5299	·5324	·5348	·5373	·5398	·5422	·5446	57	2	5	7	10	12	15	17	20	22
33	·5446	·5471	·5495	·5519	·5544	·5568	·5592	56	2	5	7	10	12	15	17	19	22
34	·5592	·5616	·5640	·5664	·5688	·5712	·5736	55	2	5	7	10	12	14	17	19	22
35	0·5736	0·5760	0·5783	0·5807	0·5831	0·5854	0·5878	54	2	5	7	9	12	14	17	19	21
36	·5878	·5901	·5925	·5948	·5972	·5995	·6018	53	2	5	7	9	12	14	16	19	21
37	·6018	·6041	·6065	·6088	·6111	·6134	·6157	52	2	5	7	9	12	14	16	18	21
38	·6157	·6180	·6202	·6225	·6248	·6271	·6293	51	2	5	7	9	11	14	16	18	20
39	·6293	·6316	·6338	·6361	·6383	·6406	·6428	50°	2	4	7	9	11	13	16	18	20
40°	0·6428	0·6450	0·6472	0·6494	0·6517	0·6539	0·6561	49	2	4	7	9	11	13	15	18	20
41	·6561	·6583	·6604	·6626	·6648	·6670	·6691	48	2	4	7	9	11	13	15	17	20
42	·6691	·6713	·6734	·6756	·6777	·6799	·6820	47	2	4	6	9	11	13	15	17	19
43	·6820	·6841	·6862	·6884	·6905	·6926	·6947	46	2	4	6	8	11	13	15	17	19
44	·6947	·6967	·6988	·7009	·7030	·7050	·7071	45	2	4	6	8	10	12	15	17	19
	60′	50′	40′	30′	20′	10′	0′		1′	2′	3′	4′	5′	6′	7′	8′	9′

COSINES.

	0′	10′	20′	30′	40′	50′	60′		Differences. 1′	2′	3′	4′	5′	6′	7′	8′	9′
45°	0·7071	0·7092	0·7112	0·7133	0·7153	0·7173	0·7193	44°	2	4	6	8	10	12	14	16	18
46	·7193	·7214	·7234	·7254	·7274	·7294	·7314	43	2	4	6	8	10	12	14	16	18
47	·7314	·7333	·7353	·7373	·7392	·7412	·7431	42	2	4	6	8	10	12	14	16	18
48	·7431	·7451	·7470	·7490	·7509	·7528	·7547	41	2	4	6	8	10	12	13	15	17
49	·7547	·7566	·7585	·7604	·7623	·7642	·7660	40°	2	4	6	8	9	11	13	15	17
50°	0·7660	0·7679	0·7698	0·7716	0·7735	0·7753	0·7771	39	2	4	6	7	9	11	13	15	17
51	·7771	·7790	·7808	·7826	·7844	·7862	·7880	38	2	4	5	7	9	11	13	14	16
52	·7880	·7898	·7916	·7934	·7951	·7969	·7986	37	2	4	5	7	9	11	12	14	16
53	·7986	·8004	·8021	·8039	·8056	·8073	·8090	36	2	3	5	7	9	10	12	14	16
54	·8090	·8107	·8124	·8141	·8158	·8175	·8192	35	2	3	5	7	8	10	12	14	15
55	0·8192	0·8208	0·8225	0·8241	0·8258	0·8274	0·8290	34	2	3	5	7	8	10	12	13	15
56	·8290	·8307	·8323	·8339	·8355	·8371	·8387	33	2	3	5	6	8	10	11	13	14
57	·8387	·8403	·8418	·8434	·8450	·8465	·8480	32	2	3	5	6	8	9	11	13	14
58	·8480	·8496	·8511	·8526	·8542	·8557	·8572	31	2	3	5	6	8	9	11	12	14
59	·8572	·8587	·8601	·8616	·8631	·8646	·8660	30°	1	3	4	6	7	9	10	12	13
60°	0·8660	0·8675	0·8689	0·8704	0·8718	0·8732	0·8746	29	1	3	4	6	7	9	10	11	13
61	·8746	·8760	·8774	·8788	·8802	·8816	·8829	28	1	3	4	6	7	8	10	11	12
62	·8829	·8843	·8857	·8870	·8884	·8897	·8910	27	1	3	4	5	7	8	9	11	12
63	·8910	·8923	·8936	·8949	·8962	·8975	·8988	26	1	3	4	5	6	8	9	10	12
64	·8988	·9001	·9013	·9026	·9038	·9051	·9063	25	1	3	4	5	6	8	9	10	11
65	0·9063	0·9075	0·9088	0·9100	0·9112	0·9124	0·9135	24	1	2	4	5	6	7	8	10	11
66	·9135	·9147	·9159	·9171	·9182	·9194	·9205	23	1	2	3	5	6	7	8	9	10
67	·9205	·9216	·9228	·9239	·9250	·9261	·9272	22	1	2	3	4	6	7	8	9	10
68	·9272	·9283	·9293	·9304	·9315	·9325	·9336	21	1	2	3	4	5	6	7	9	10
69	·9336	·9346	·9356	·9367	·9377	·9387	·9397	20°	1	2	3	4	5	6	7	8	9
70°	0·9397	0·9407	0·9417	0·9426	0·9436	0·9446	0·9455	19	1	2	3	4	5	6	7	8	9
71	·9455	·9465	·9474	·9483	·9492	·9502	·9511	18	1	2	3	4	5	6	6	7	8
72	·9511	·9520	·9528	·9537	·9546	·9555	·9563	17	1	2	3	4	4	5	6	7	8
73	·9563	·9572	·9580	·9588	·9596	·9605	·9613	16	1	2	2	3	4	5	6	7	7
74	·9613	·9621	·9628	·9636	·9644	·9652	·9659	15	1	2	2	3	4	5	5	6	7
75	0·9659	0·9667	0·9674	0·9681	0·9689	0·9696	0·9703	14	1	1	2	3	4	4	5	6	7
76	·9703	·9710	·9717	·9724	·9730	·9737	·9744	13	1	1	2	3	3	4	5	5	6
77	·9744	·9750	·9757	·9763	·9769	·9775	·9781	12	1	1	2	3	3	4	4	5	6
78	·9781	·9787	·9793	·9799	·9805	·9811	·9816	11	1	1	2	2	3	3	4	5	5
79	·9816	·9822	·9827	·9833	·9838	·9843	·9848	10°	1	1	2	2	3	3	4	4	5
80°	0·9848	0·9853	0·9858	0·9863	0·9868	0·9872	0·9877	9	0	1	1	2	2	3	3	4	4
81	·9877	·9881	·9886	·9890	·9894	·9899	·9903	8	0	1	1	2	2	3	3	3	4
82	·9903	·9907	·9911	·9914	·9918	·9922	·9925	7	0	1	1	2	2	2	3	3	3
83	·9925	·9929	·9932	·9936	·9939	·9942	·9945	6	0	1	1	1	2	2	2	3	3
84	·9945	·9948	·9951	·9954	·9957	·9959	·9962	5	0	1	1	1	1	2	2	2	3
85	0·9962	0·9964	0·9967	0·9969	0·9971	0·9974	0·9976	4									
86	·9976	·9978	·9980	·9981	·9983	·9985	·9986	3									
87	·9986	·9988	·9989	·9990	·9992	·9993	·9994	2									
88	·9994	·9995	·9996	·9997	·9997	·9998	·9998	1									
89	0·9998	0·9999	0·9999	1·0000	1·0000	1·0000	1·0000	0°									
	60′	50′	40′	30′	20′	10′	0′		1′	2′	3′	4′	5′	6′	7′	8′	9′

Differences are so small here that tabulation is unnecessary.

COSINES.

	0′	10′	20′	30′	40′	50′	60′		1′	2′	3′	4′	5′	6′	7′	8′	9′
									Differences.								
0°	0·0000	0·0029	0·0058	0·0087	0·0116	0·0145	0·0175	89°	3	6	9	12	15	17	20	23	26
1	·0175	·0204	·0233	·0262	·0291	·0320	·0349	88	3	6	9	12	15	17	20	23	26
2	·0349	·0378	·0407	·0437	·0466	·0495	·0524	87	3	6	9	12	15	18	20	23	26
3	·0524	·0553	·0582	·0612	·0641	·0670	·0699	86	3	6	9	12	15	18	20	23	26
4	·0699	·0729	·0758	·0787	·0816	·0846	·0875	85	3	6	9	12	15	18	21	23	26
5	0·0875	0·0904	0·0934	0·0963	0·0992	0·1022	0·1051	84	3	6	9	12	15	18	21	24	26
6	·1051	·1080	·1110	·1139	·1169	·1198	·1228	83	3	6	9	12	15	18	21	24	27
7	·1228	·1257	·1287	·1317	·1346	·1376	·1405	82	3	6	9	12	15	18	21	24	27
8	·1405	·1435	·1465	·1495	·1524	·1554	·1584	81	3	6	9	12	15	18	21	24	27
9	·1584	·1614	·1644	·1673	·1703	·1733	·1763	80°	3	6	9	12	15	18	21	24	27
10°	0·1763	0·1793	0·1823	0·1853	0·1883	0·1914	0·1944	79	3	6	9	12	15	18	21	24	27
11	·1944	·1974	·2004	·2035	·2065	·2095	·2126	78	3	6	9	12	15	18	21	24	27
12	·2126	·2156	·2186	·2217	·2247	·2278	·2309	77	3	6	9	12	15	18	21	24	27
13	·2309	·2339	·2370	·2401	·2432	·2462	·2493	76	3	6	9	12	15	18	22	25	28
14	·2493	·2524	·2555	·2586	·2617	·2648	·2679	75	3	6	9	12	16	19	22	25	28
15	0·2679	0·2711	0·2742	0·2773	0·2805	0·2836	0·2867	74	3	6	9	13	16	19	22	25	28
16	·2867	·2899	·2931	·2962	·2994	·3026	·3057	73	3	6	9	13	16	19	22	25	28
17	·3057	·3089	·3121	·3153	·3185	·3217	·3249	72	3	6	10	13	16	19	22	26	29
18	·3249	·3281	·3314	·3346	·3378	·3411	·3443	71	3	6	10	13	16	19	23	26	29
19	·3443	·3476	·3508	·3541	·3574	·3607	·3640	70°	3	7	10	13	16	20	23	26	29
20°	0·3640	0·3673	0·3706	0·3739	0·3772	0·3805	0·3839	69	3	7	10	13	17	20	23	27	30
21	·3839	·3872	·3906	·3939	·3973	·4006	·4040	68	3	7	10	13	17	20	24	27	30
22	·4040	·4074	·4108	·4142	·4176	·4210	·4245	67	3	7	10	14	17	20	24	27	31
23	·4245	·4279	·4314	·4348	·4383	·4417	·4452	66	3	7	10	14	17	21	24	28	31
24	·4452	·4487	·4522	·4557	·4592	·4628	·4663	65	4	7	11	14	18	21	25	28	32
25	0·4663	0·4699	0·4734	0·4770	0·4806	0·4841	0·4877	64	4	7	11	14	18	21	25	29	32
26	·4877	·4913	·4950	·4986	·5022	·5059	·5095	63	4	7	11	15	18	22	25	29	33
27	·5095	·5132	·5169	·5206	·5243	·5280	·5317	62	4	7	11	15	18	22	26	30	33
28	·5317	·5354	·5392	·5430	·5467	·5505	·5543	61	4	8	11	15	19	23	26	30	34
29	·5543	·5581	·5619	·5658	·5696	·5735	·5774	60°	4	8	12	15	19	23	27	31	35
30°	0·5774	0·5812	0·5851	0·5890	0·5930	0·5969	0·6009	59	4	8	12	16	20	24	27	31	35
31	·6009	·6048	·6088	·6128	·6168	·6208	·6249	58	4	8	12	16	20	24	28	32	36
32	·6249	·6289	·6330	·6371	·6412	·6453	·6494	57	4	8	12	16	20	25	29	33	37
33	·6494	·6536	·6577	·6619	·6661	·6703	·6745	56	4	8	13	17	21	25	29	33	38
34	·6745	·6787	·6830	·6873	·6916	·6959	·7002	55	4	9	13	17	21	26	30	34	39
35	0·7002	0·7046	0·7089	0·7133	0·7177	0·7221	0·7265	54	4	9	13	18	22	26	31	35	40
36	·7265	·7310	·7355	·7400	·7445	·7490	·7536	53	5	9	14	18	23	27	32	36	41
37	·7536	·7581	·7627	·7673	·7720	·7766	·7813	52	5	9	14	18	23	28	32	37	42
38	·7813	·7860	·7907	·7954	·8002	·8050	·8098	51	5	10	14	19	24	29	33	38	43
39	·8098	·8146	·8195	·8243	·8292	·8342	·8391	50°	5	10	15	20	24	29	34	39	44
40°	0·8391	0·8441	0·8491	0·8541	0·8591	0·8642	0·8693	49	5	10	15	20	25	30	35	40	45
41	·8693	·8744	·8796	·8847	·8899	·8952	·9004	48	5	10	16	21	26	31	36	41	47
42	·9004	·9057	·9110	·9163	·9217	·9271	·9325	47	5	11	16	21	27	32	37	43	48
43	·9325	·9380	·9435	·9490	·9545	·9601	·9657	46	6	11	17	22	28	33	39	44	50
44	·9657	·9713	·9770	·9827	·9884	·9942	1·0000	45	6	11	17	23	29	34	40	46	51
	60′	50′	40′	30′	20′	10′	0′		1′	2′	3′	4′	5′	6′	7′	8′	9′

COTANGENTS.

	0′	10′	20′	30′	40′	50′	60′		Differences. 1′	2′	3′	4′	5′	6′	7′	8′	9′
45°	1·0000	1·0058	1·0117	1·0176	1·0235	1·0295	1·0355	44°	6	12	18	24	30	36	41	47	53
46	·0355	·0416	·0477	·0538	·0599	·0661	·0724	43	6	12	18	25	31	37	43	49	55
47	·0724	·0786	·0850	·0913	·0977	·1041	·1106	42	6	13	19	26	32	38	45	51	57
48	·1106	·1171	·1237	·1303	·1369	·1436	·1504	41	7	13	20	27	33	40	46	53	60
49	·1504	·1571	·1640	·1708	·1778	·1847	·1918	40°	7	14	21	28	34	41	48	55	62
50°	1·1918	1·1988	1·2059	1·2131	1·2203	1·2276	1·2349	39	7	14	22	29	36	43	50	58	65
51	·2349	·2423	·2497	·2572	·2647	·2723	·2799	38	8	15	23	30	38	45	53	60	68
52	·2799	·2876	·2954	·3032	·3111	·3190	·3270	37	8	16	24	31	39	47	55	63	71
53	·3270	·3351	·3432	·3514	·3597	·3680	·3764	36	8	16	25	33	41	49	58	66	74
54	·3764	·3848	·3934	·4019	·4106	·4193	·4281	35	9	17	26	35	43	52	60	69	78
55	1·4281	1·4370	1·4460	1·4550	1·4641	1·4733	1·4826	34	9	18	27	36	45	54	63	73	82
56	·4826	·4919	·5013	·5108	·5204	·5301	·5399	33	10	19	29	38	48	57	67	76	86
57	·5399	·5497	·5597	·5697	·5798	·5900	·6003	32	10	20	30	40	50	60	71	81	91
58	·6003	·6107	·6212	·6319	·6426	·6534	·6643	31	11	21	32	43	53	64	75	85	96
59	·6643	·6753	·6864	·6977	·7090	·7205	·7321	30°	11	23	34	45	56	68	79	90	102
60°	1·732	1·744	1·756	1·767	1·780	1·792	1·804	29	1	2	4	5	6	7	8	10	11
61	1·804	1·816	1·829	1·842	1·855	1·868	1·881	28	1	3	4	5	6	8	9	10	12
62	1·881	1·894	1·907	1·921	1·935	1·949	1·963	27	1	3	4	5	7	8	10	11	12
63	1·963	1·977	1·991	2·006	2·020	2·035	2·050	26	1	3	4	6	7	9	10	12	13
64	2·050	2·066	2·081	2·097	2·112	2·128	2·145	25	2	3	5	6	8	9	11	13	14
65	2·145	2·161	2·177	2·194	2·211	2·229	2·246	24	2	3	5	7	8	10	12	14	15
66	2·246	2·264	2·282	2·300	2·318	2·337	2·356	23	2	4	5	7	9	11	13	15	16
67	2·356	2·375	2·394	2·414	2·434	2·455	2·475	22	2	4	6	8	10	12	14	16	18
68	2·475	2·496	2·517	2·539	2·560	2·583	2·605	21	2	4	6	9	11	13	15	17	20
69	2·605	2·628	2·651	2·675	2·699	2·723	2·747	20°	2	5	7	9	12	14	17	19	21
70°	2·747	2·773	2·798	2·824	2·850	2·877	2·904	19	3	5	8	10	13	16	18	21	23
71	2·904	2·932	2·960	2·989	3·018	3·047	3·078	18	3	6	9	12	14	17	20	23	26
72	3·078	3·108	3·140	3·172	3·204	3·237	3·271	17	3	6	10	13	16	19	23	26	29
73	3·271	3·305	3·340	3·376	3·412	3·450	3·487	16	4	7	11	14	18	22	25	29	32
74	3·487	3·526	3·566	3·606	3·647	3·689	3·732	15	4	8	12	16	20	24	29	33	37
75	3·732	3·776	3·821	3·867	3·914	3·962	4·011	14	5	9	14	19	23	28	33	37	42
76	4·011	4·061	4·113	4·165	4·219	4·275	4·331	13	5	11	16	21	27	32	37	43	48
77	4·331	4·390	4·449	4·511	4·574	4·638	4·705	12	6	12	19	25	31	37	44	50	56
78	4·705	4·773	4·843	4·915	4·989	5·066	5·145	11	7	15	22	29	37	44	51	59	66
79	5·145	5·226	5·309	5·396	5·485	5·576	5·671	10°	9	18	26	35	44	53	61	70	79
80°	5·671	5·769	5·871	5·976	6·084	6·197	6·314	9									
81	6·314	6·435	6·561	6·691	6·827	6·968	7·115	8									
82	7·115	7·269	7·429	7·596	7·770	7·953	8·144	7									
83	8·144	8·345	8·556	8·777	9·010	9·255	9·514	6									
84	9·514	9·788	10·078	10·385	10·712	11·059	11·430	5									
85	11·43	11·83	12·25	12·71	13·20	13·73	14·30	4									
86	14·30	14·92	15·60	16·35	17·17	18·07	19·08	3									
87	19·08	20·21	21·47	22·90	24·54	26·43	28·64	2									
88	28·64	31·24	34·37	38·19	42·96	49·10	57·29	1									
89	57·29	68·75	85·94	114·59	171·89	343·77	∞	0°									
	60′	50′	40′	30′	20′	10′	0′		1′	2′	3′	4′	5′	6′	7′	8′	9′

The differences change so rapidly here that they cannot be tabulated.

The cotangent of a small angle of *n* minutes of arc or the tangent of 90° minus *n* minutes is very nearly equal to 3438 divided by *n*.

COTANGENTS.

For small angles of n minutes of arc, log sine n' or log cosine $(90° - n')$ $= \log n + \bar{4}·4637$

Differences vary so rapidly here that tabulation is impossible.

	0′	10′	20′	30′	40′	50′	60′		1′	2′	3′	4′	5′	6′	7′	8′	9′
0°	−∞	$\bar{3}$·4637	$\bar{3}$·7648	$\bar{3}$·9408	$\bar{2}$·0658	$\bar{2}$·1627	$\bar{2}$·2419	89°									
1	$\bar{2}$·2419	$\bar{2}$·3088	$\bar{2}$·3668	$\bar{2}$·4179	$\bar{2}$·4637	$\bar{2}$·5050	$\bar{2}$·5428	88									
2	$\bar{2}$·5428	$\bar{2}$·5776	$\bar{2}$·6097	$\bar{2}$·6397	$\bar{2}$·6677	$\bar{2}$·6940	$\bar{2}$·7188	87									
3	$\bar{2}$·7188	$\bar{2}$·7423	$\bar{2}$·7645	$\bar{2}$·7857	$\bar{2}$·8059	$\bar{2}$·8251	$\bar{2}$·8436	86									
4	$\bar{2}$·8436	$\bar{2}$·8613	$\bar{2}$·8783	$\bar{2}$·8946	$\bar{2}$·9104	$\bar{2}$·9256	$\bar{2}$·9403	85									
5	$\bar{2}$·9403	$\bar{2}$·9545	$\bar{2}$·9682	$\bar{2}$·9816	$\bar{2}$·9945	$\bar{1}$·0070	$\bar{1}$·0192	84									
6	$\bar{1}$·0192	$\bar{1}$·0311	$\bar{1}$·0426	$\bar{1}$·0539	$\bar{1}$·0648	$\bar{1}$·0755	$\bar{1}$·0859	83									
7	$\bar{1}$·0859	$\bar{1}$·0961	$\bar{1}$·1060	$\bar{1}$·1157	$\bar{1}$·1252	$\bar{1}$·1345	$\bar{1}$·1436	82	10	19	29	39	48	58	67	77	87
8	$\bar{1}$·1436	$\bar{1}$·1525	$\bar{1}$·1612	$\bar{1}$·1697	$\bar{1}$·1781	$\bar{1}$·1863	$\bar{1}$·1943	81	8	17	25	34	42	51	59	68	76
9	$\bar{1}$·1943	$\bar{1}$·2022	$\bar{1}$·2100	$\bar{1}$·2176	$\bar{1}$·2251	$\bar{1}$·2324	$\bar{1}$·2397	80°	8	15	23	30	38	45	53	61	68
10°	$\bar{1}$·2397	$\bar{1}$·2468	$\bar{1}$·2538	$\bar{1}$·2606	$\bar{1}$·2674	$\bar{1}$·2740	$\bar{1}$·2806	79	7	14	20	27	34	41	48	55	62
11	·2806	·2870	·2934	·2997	·3058	·3119	·3179	78	6	12	19	25	31	37	44	50	56
12	·3179	·3238	·3296	·3353	·3410	·3466	·3521	77	6	11	17	23	29	34	40	46	51
13	·3521	·3575	·3629	·3682	·3734	·3786	·3837	76	5	11	16	21	26	32	37	42	47
14	·3837	·3887	·3937	·3986	·4035	·4083	·4130	75	5	10	15	20	24	29	34	39	44
15	$\bar{1}$·4130	$\bar{1}$·4177	$\bar{1}$·4223	$\bar{1}$·4269	$\bar{1}$·4314	$\bar{1}$·4359	$\bar{1}$·4403	74	5	9	14	18	23	27	32	36	41
16	·4403	·4447	·4491	·4533	·4576	·4618	·4659	73	4	9	13	17	21	26	30	34	38
17	·4659	·4700	·4741	·4781	·4821	·4861	·4900	72	4	8	12	16	20	24	28	32	36
18	·4900	·4939	·4977	·5015	·5052	·5090	·5126	71	4	8	11	15	19	23	26	30	34
19	·5126	·5163	·5199	·5235	·5270	·5306	·5341	70°	4	7	11	14	18	21	25	29	32
20°	$\bar{1}$·5341	$\bar{1}$·5375	$\bar{1}$·5409	$\bar{1}$·5443	$\bar{1}$·5477	$\bar{1}$·5510	$\bar{1}$·5543	69	3	7	10	14	17	20	24	27	30
21	·5543	·5576	·5609	·5641	·5673	·5704	·5736	68	3	6	10	13	16	19	22	26	29
22	·5736	·5767	·5798	·5828	·5859	·5889	·5919	67	3	6	9	12	15	18	21	24	27
23	·5919	·5948	·5978	·6007	·6036	·6065	·6093	66	3	6	9	12	15	17	20	23	26
24	·6093	·6121	·6149	·6177	·6205	·6232	·6259	65	3	6	8	11	14	17	19	22	25
25	$\bar{1}$·6259	$\bar{1}$·6286	$\bar{1}$·6313	$\bar{1}$·6340	$\bar{1}$·6366	$\bar{1}$·6392	$\bar{1}$·6418	64	3	5	8	11	13	16	19	21	24
26	·6418	·6444	·6470	·6495	·6521	·6546	·6570	63	3	5	8	10	13	15	18	20	23
27	·6570	·6595	·6620	·6644	·6668	·6692	·6716	62	2	5	7	10	12	15	17	19	22
28	·6716	·6740	·6763	·6787	·6810	·6833	·6856	61	2	5	7	9	12	14	16	19	21
29	·6856	·6878	·6901	·6923	·6946	·6968	·6990	60°	2	4	7	9	11	13	16	18	20
30°	$\bar{1}$·6990	$\bar{1}$·7012	$\bar{1}$·7033	$\bar{1}$·7055	$\bar{1}$·7076	$\bar{1}$·7097	$\bar{1}$·7118	59	2	4	6	9	11	13	15	17	19
31	·7118	·7139	·7160	·7181	·7201	·7222	·7242	58	2	4	6	8	10	12	14	16	19
32	·7242	·7262	·7282	·7302	·7322	·7342	·7361	57	2	4	6	8	10	12	14	16	18
33	·7361	·7380	·7400	·7419	·7438	·7457	·7476	56	2	4	6	8	10	11	13	15	17
34	·7476	·7494	·7513	·7531	·7550	·7568	·7586	55	2	4	6	7	9	11	13	15	17
35	$\bar{1}$·7586	$\bar{1}$·7604	$\bar{1}$·7622	$\bar{1}$·7640	$\bar{1}$·7657	$\bar{1}$·7675	$\bar{1}$·7692	54	2	4	5	7	9	11	12	14	16
36	·7692	·7710	·7727	·7744	·7761	·7778	·7795	53	2	3	5	7	9	10	12	14	15
37	·7795	·7811	·7828	·7844	·7861	·7877	·7893	52	2	3	5	7	8	10	12	13	15
38	·7893	·7910	·7926	·7941	·7957	·7973	·7989	51	2	3	5	6	8	10	11	13	14
39	·7989	·8004	·8020	·8035	·8050	·8066	·8081	50°	2	3	5	6	8	9	11	12	14
40°	$\bar{1}$·8081	$\bar{1}$·8096	$\bar{1}$·8111	$\bar{1}$·8125	$\bar{1}$·8140	$\bar{1}$·8155	$\bar{1}$·8169	49	1	3	4	6	7	9	10	12	13
41	·8169	·8184	·8198	·8213	·8227	·8241	·8255	48	1	3	4	6	7	9	10	11	13
42	·8255	·8269	·8283	·8297	·8311	·8324	·8338	47	1	3	4	6	7	8	10	11	12
43	·8338	·8351	·8365	·8378	·8391	·8405	·8418	46	1	3	4	5	7	8	9	11	12
44	·8418	·8431	·8444	·8457	·8469	·8482	·8495	45	1	3	4	5	6	8	9	10	12
	60′	50′	40′	30′	20′	10′	0′		1′	2′	3′	4′	5′	6′	7′	8′	9′

LOGARITHMS OF COSINES.

	0′	10′	20′	30′	40′	50′	60′		1′	2′	3′	4′	5′	6′	7′	8′	9′
									Differences.								
5°	Ī·8495	Ī·8507	Ī·8520	Ī·8532	Ī·8545	Ī·8557	Ī·8569	44°	1	2	4	5	6	7	9	10	11
6	·8569	·8582	·8594	·8606	·8618	·8629	·8641	43	1	2	4	5	6	7	8	10	11
7	·8641	·8653	·8665	·8676	·8688	·8699	·8711	42	1	2	3	5	6	7	8	9	10
8	·8711	·8722	·8733	·8745	·8756	·8767	·8778	41	1	2	3	4	6	7	8	9	10
9	·8778	·8789	·8800	·8810	·8821	·8832	·8843	40°	1	2	3	4	5	6	8	9	10
0°	Ī·8843	Ī·8853	Ī·8864	Ī·8874	Ī·8884	Ī·8895	Ī·8905	39	1	2	3	4	5	6	7	8	9
1	·8905	·8915	·8925	·8935	·8945	·8955	·8965	38	1	2	3	4	5	6	7	8	9
2	·8965	·8975	·8985	·8995	·9004	·9014	·9023	37	1	2	3	4	5	6	7	8	9
3	·9023	·9033	·9042	·9052	·9061	·9070	·9080	36	1	2	3	4	5	6	7	7	8
4	·9080	·9089	·9098	·9107	·9116	·9125	·9134	35	1	2	3	4	5	5	6	7	8
5	Ī·9134	Ī·9142	Ī·9151	Ī·9160	Ī·9169	Ī·9177	Ī·9186	34	1	2	3	3	4	5	6	7	8
6	·9186	·9194	·9203	·9211	·9219	·9228	·9236	33	1	2	3	3	4	5	6	7	8
7	·9236	·9244	·9252	·9260	·9268	·9276	·9284	32	1	2	2	3	4	5	6	6	7
8	·9284	·9292	·9300	·9308	·9315	·9323	·9331	31	1	2	2	3	4	5	5	6	7
9	·9331	·9338	·9346	·9353	·9361	·9368	·9375	30°	1	1	2	3	4	4	5	6	7
0°	Ī·9375	Ī·9383	Ī·9390	Ī·9397	Ī·9404	Ī·9411	Ī·9418	29	1	1	2	3	4	4	5	6	6
1	·9418	·9425	·9432	·9439	·9446	·9453	·9459	28	1	1	2	3	3	4	5	5	6
2	·9459	·9466	·9473	·9479	·9486	·9492	·9499	27	1	1	2	3	3	4	5	5	6
3	·9499	·9505	·9512	·9518	·9524	·9530	·9537	26	1	1	2	3	3	4	4	5	6
4	·9537	·9543	·9549	·9555	·9561	·9567	·9573	25	1	1	2	2	3	4	4	5	5
5	Ī·9573	Ī·9579	Ī·9584	Ī·9590	Ī·9596	Ī·9602	Ī·9607	24	1	1	2	2	3	3	4	5	5
6	·9607	·9613	·9618	·9624	·9629	·9635	·9640	23	1	1	2	2	3	3	4	4	5
7	·9640	·9646	·9651	·9656	·9661	·9667	·9672	22	1	1	2	2	3	3	4	4	5
8	·9672	·9677	·9682	·9687	·9692	·9697	·9702	21	0	1	1	2	2	3	3	4	4
9	·9702	·9706	·9711	·9716	·9721	·9725	·9730	20°	0	1	1	2	2	3	3	4	4
0°	Ī·9730	Ī·9734	Ī·9739	Ī·9743	Ī·9748	Ī·9752	Ī·9757	19	0	1	1	2	2	3	3	4	4
1	·9757	·9761	·9765	·9770	·9774	·9778	·9782	18	0	1	1	2	2	3	3	3	4
2	·9782	·9786	·9790	·9794	·9798	·9802	·9806	17	0	1	1	2	2	2	3	3	4
3	·9806	·9810	·9814	·9817	·9821	·9825	·9828	16	0	1	1	1	2	2	3	3	3
4	·9828	·9832	·9836	·9839	·9843	·9846	·9849	15	0	1	1	1	2	2	2	3	3
5	Ī·9849	Ī·9853	Ī·9856	Ī·9859	Ī·9863	Ī·9866	Ī·9869	14									
6	·9869	·9872	·9875	·9878	·9881	·9884	·9887	13									
7	·9887	·9890	·9893	·9896	·9899	·9901	·9904	12									
8	·9904	·9907	·9909	·9912	·9914	·9917	·9919	11									
9	·9919	·9922	·9924	·9927	·9929	·9931	·9934	10°									
0°	Ī·9934	Ī·9936	Ī·9938	Ī·9940	Ī·9942	Ī·9944	Ī·9946	9									
1	·9946	·9948	·9950	·9952	·9954	·9956	·9958	8									
2	·9958	·9959	·9961	·9963	·9964	·9966	·9968	7									
3	·9968	·9969	·9971	·9972	·9973	·9975	·9976	6									
4	·9976	·9977	·9979	·9980	·9981	·9982	·9983	5									
5	Ī·9983	Ī·9985	Ī·9986	Ī·9987	Ī·9988	Ī·9989	Ī·9989	4									
6	·9989	·9990	·9991	·9992	·9993	·9993	·9994	3									
7	·9994	·9995	·9995	·9996	·9996	·9997	·9997	2									
8	·9997	·9998	·9998	·9999	·9999	·9999	·9999	1									
9	Ī·9999	0·0000	0·0000	0·0000	0·0000	0·0000	0·0000	0°									
	60′	50′	40′	30′	20′	10′	0′		1′	2′	3′	4′	5′	6′	7′	8′	9′

Differences are so small here that tabulation is unnecessary.

LOGARITHMS OF COSINES.

	0′	10′	20′	30′	40′	50′	60′		Differences. 1′	2′	3′	4′	5′	6′	7′	8′	9′
0°	−∞	$\bar{3}$·4637	$\bar{3}$·7648	$\bar{3}$·9409	$\bar{2}$·0658	$\bar{2}$·1627	$\bar{2}$·2419	89°									
1	$\bar{2}$·2419	$\bar{2}$·3089	$\bar{2}$·3669	$\bar{2}$·4181	$\bar{2}$·4638	$\bar{2}$·5053	$\bar{2}$·5431	88									
2	$\bar{2}$·5431	$\bar{2}$·5779	$\bar{2}$·6101	$\bar{2}$·6401	$\bar{2}$·6682	$\bar{2}$·6945	$\bar{2}$·7194	87									
3	$\bar{2}$·7194	$\bar{2}$·7429	$\bar{2}$·7652	$\bar{2}$·7865	$\bar{2}$·8067	$\bar{2}$·8261	$\bar{2}$·8446	86									
4	$\bar{2}$·8446	$\bar{2}$·8624	$\bar{2}$·8795	$\bar{2}$·8960	$\bar{2}$·9118	$\bar{2}$·9272	$\bar{2}$·9420	85									
5	$\bar{2}$·9420	$\bar{2}$·9563	$\bar{2}$·9701	$\bar{2}$·9836	$\bar{2}$·9966	$\bar{1}$·0093	$\bar{1}$·0216	84									
6	$\bar{1}$·0216	$\bar{1}$·0336	$\bar{1}$·0453	$\bar{1}$·0567	$\bar{1}$·0678	$\bar{1}$·0786	$\bar{1}$·0891	83									
7	$\bar{1}$·0891	$\bar{1}$·0995	$\bar{1}$·1096	$\bar{1}$·1194	$\bar{1}$·1291	$\bar{1}$·1385	$\bar{1}$·1478	82	10	20	29	39	49	59	69	78	88
8	$\bar{1}$·1478	$\bar{1}$·1569	$\bar{1}$·1658	$\bar{1}$·1745	$\bar{1}$·1831	$\bar{1}$·1915	$\bar{1}$·1997	81	9	17	26	35	43	52	61	69	78
9	$\bar{1}$·1997	$\bar{1}$·2078	$\bar{1}$·2158	$\bar{1}$·2236	$\bar{1}$·2313	$\bar{1}$·2389	$\bar{1}$·2463	80°	8	16	23	31	39	47	54	62	70
10°	$\bar{1}$·2463	$\bar{1}$·2536	$\bar{1}$·2609	$\bar{1}$·2680	$\bar{1}$·2750	$\bar{1}$·2819	$\bar{1}$·2887	79	7	14	21	28	35	42	49	57	64
11	·2887	·2953	·3020	·3085	·3149	·3212	·3275	78	6	13	19	26	32	39	45	52	58
12	·3275	·3336	·3397	·3458	·3517	·3576	·3634	77	6	12	18	24	30	36	42	48	54
13	·3634	·3691	·3748	·3804	·3859	·3914	·3968	76	6	11	17	22	28	33	39	45	50
14	·3968	·4021	·4074	·4127	·4178	·4230	·4281	75	5	10	16	21	26	31	37	42	47
15	$\bar{1}$·4281	$\bar{1}$·4331	$\bar{1}$·4381	$\bar{1}$·4430	$\bar{1}$·4479	$\bar{1}$·4527	$\bar{1}$·4575	74	5	10	15	20	25	30	34	39	44
16	·4575	·4622	·4669	·4716	·4762	·4808	·4853	73	5	9	14	19	23	28	33	37	42
17	·4853	·4898	·4943	·4987	·5031	·5075	·5118	72	4	9	13	18	22	26	31	35	40
18	·5118	·5161	·5203	·5245	·5287	·5329	·5370	71	4	8	13	17	21	25	29	34	38
19	·5370	·5411	·5451	·5491	·5531	·5571	·5611	70°	4	8	12	16	20	24	28	32	36
20°	$\bar{1}$·5611	$\bar{1}$·5650	$\bar{1}$·5689	$\bar{1}$·5727	$\bar{1}$·5766	$\bar{1}$·5804	$\bar{1}$·5842	69	4	8	12	15	19	23	27	31	35
21	·5842	·5879	·5917	·5954	·5991	·6028	·6064	68	4	7	11	15	19	22	26	30	33
22	·6064	·6100	·6136	·6172	·6208	·6243	·6279	67	4	7	11	14	18	21	25	29	32
23	·6279	·6314	·6348	·6383	·6417	·6452	·6486	66	3	7	10	14	17	21	24	28	31
24	·6486	·6520	·6553	·6587	·6620	·6654	·6687	65	3	7	10	13	17	20	23	27	30
25	$\bar{1}$·6687	$\bar{1}$·6720	$\bar{1}$·6752	$\bar{1}$·6785	$\bar{1}$·6817	$\bar{1}$·6850	$\bar{1}$·6882	64	3	7	10	13	16	20	23	26	29
26	·6882	·6914	·6946	·6977	·7009	·7040	·7072	63	3	6	9	13	16	19	22	25	28
27	·7072	·7103	·7134	·7165	·7196	·7226	·7257	62	3	6	9	12	15	19	22	25	28
28	·7257	·7287	·7317	·7348	·7378	·7408	·7438	61	3	6	9	12	15	18	21	24	27
29	·7438	·7467	·7497	·7526	·7556	·7585	·7614	60°	3	6	9	12	15	18	21	24	27
30°	$\bar{1}$·7614	$\bar{1}$·7644	$\bar{1}$·7673	$\bar{1}$·7701	$\bar{1}$·7730	$\bar{1}$·7759	$\bar{1}$·7788	59	3	6	9	12	14	17	20	23	26
31	·7788	·7816	·7845	·7873	·7902	·7930	·7958	58	3	6	9	11	14	17	20	23	26
32	·7958	·7986	·8014	·8042	·8070	·8097	·8125	57	3	6	8	11	14	17	20	22	25
33	·8125	·8153	·8180	·8208	·8235	·8263	·8290	56	3	5	8	11	14	16	19	22	25
34	·8290	·8317	·8344	·8371	·8398	·8425	·8452	55	3	5	8	11	14	16	19	22	24
35	$\bar{1}$·8452	$\bar{1}$·8479	$\bar{1}$·8506	$\bar{1}$·8533	$\bar{1}$·8559	$\bar{1}$·8586	$\bar{1}$·8613	54	3	5	8	11	13	16	19	21	24
36	·8613	·8639	·8666	·8692	·8718	·8745	·8771	53	3	5	8	11	13	16	19	21	24
37	·8771	·8797	·8824	·8850	·8876	·8902	·8928	52	3	5	8	10	13	16	18	21	24
38	·8928	·8954	·8980	·9006	·9032	·9058	·9084	51	3	5	8	10	13	16	18	21	23
39	·9084	·9110	·9135	·9161	·9187	·9212	·9238	50°	3	5	8	10	13	15	18	21	23
40°	$\bar{1}$·9238	$\bar{1}$·9264	$\bar{1}$·9289	$\bar{1}$·9315	$\bar{1}$·9341	$\bar{1}$·9366	$\bar{1}$·9392	49	3	5	8	10	13	15	18	20	23
41	·9392	·9417	·9443	·9468	·9494	·9519	·9544	48	3	5	8	10	13	15	18	20	23
42	·9544	·9570	·9595	·9621	·9646	·9671	·9697	47	3	5	8	10	13	15	18	20	23
43	·9697	·9722	·9747	·9773	·9798	·9823	·9848	46	3	5	8	10	13	15	18	20	23
44	·9848	·9874	·9899	·9924	·9949	·9975	0·0000	45	3	5	8	10	13	15	18	20	23
	60′	50′	40′	30′	20′	10′	0′		1′	2′	3′	4′	5′	6′	7′	8′	9′

For small angles of n minutes of arc, log tan n'
or log cot $(90° - n')$
$= \log n + \bar{4}{\cdot}4637$

Differences vary so rapidly here that tabulation is impossible.

LOGARITHMS OF COTANGENTS.

	0′	10′	20′	30′	40′	50′	60′		1′	2′	3′	4′	5′	6′	7′	8′	9′
									Differences.								
45°	0·0000	0·0025	0·0051	0·0076	0·0101	0·0126	0·0152	44°	3	5	8	10	13	15	18	20	23
46	·0152	·0177	·0202	·0228	·0253	·0278	·0303	43	3	5	8	10	13	15	18	20	23
47	·0303	·0329	·0354	·0379	·0405	·0430	·0456	42	3	5	8	10	13	15	18	20	23
48	·0456	·0481	·0506	·0532	·0557	·0583	·0608	41	3	5	8	10	13	15	18	20	23
49	·0608	·0634	·0659	·0685	·0711	·0736	·0762	40°	3	5	8	10	13	15	18	20	23
50°	0·0762	0·0788	0·0813	0·0839	0·0865	0·0890	0·0916	39	3	5	8	10	13	15	18	21	23
51	·0916	·0942	·0968	·0994	·1020	·1046	·1072	38	3	5	8	10	13	16	18	21	23
52	·1072	·1098	·1124	·1150	·1176	·1203	·1229	37	3	5	8	10	13	16	18	21	24
53	·1229	·1255	·1282	·1308	·1334	·1361	·1387	36	3	5	8	11	13	16	19	21	24
54	·1387	·1414	·1441	·1467	·1494	·1521	·1548	35	3	5	8	11	13	16	19	21	24
55	0·1548	0·1575	0·1602	0·1629	0·1656	0·1683	0·1710	34	3	5	8	11	14	16	19	22	24
56	·1710	·1737	·1765	·1792	·1820	·1847	·1875	33	3	5	8	11	14	16	19	22	25
57	·1875	·1903	·1930	·1958	·1986	·2014	·2042	32	3	6	8	11	14	17	20	22	25
58	·2042	·2070	·2098	·2127	·2155	·2184	·2212	31	3	6	9	11	14	17	20	23	26
59	·2212	·2241	·2270	·2299	·2327	·2356	·2386	30°	3	6	9	12	14	17	20	23	26
60°	0·2386	0·2415	0·2444	0·2474	0·2503	0·2533	0·2562	29	3	6	9	12	15	18	21	24	27
61	·2562	·2592	·2622	·2652	·2683	·2713	·2743	28	3	6	9	12	15	18	21	24	27
62	·2743	·2774	·2804	·2835	·2866	·2897	·2928	27	3	6	9	12	15	19	22	25	28
63	·2928	·2960	·2991	·3023	·3054	·3086	·3118	26	3	6	9	13	16	19	22	25	28
64	·3118	·3150	·3183	·3215	·3248	·3280	·3313	25	3	7	10	13	16	20	23	26	29
65	0·3313	0·3346	0·3380	0·3413	0·3447	0·3480	0·3514	24	3	7	10	13	17	20	23	27	30
66	·3514	·3548	·3583	·3617	·3652	·3686	·3721	23	3	7	10	14	17	21	24	28	31
67	·3721	·3757	·3792	·3828	·3864	·3900	·3936	22	4	7	11	14	18	21	25	29	32
68	·3936	·3972	·4009	·4046	·4083	·4121	·4158	21	4	7	11	15	19	22	26	30	33
69	·4158	·4196	·4234	·4273	·4311	·4350	·4389	20°	4	8	12	15	19	23	27	31	35
70°	0·4389	0·4429	0·4469	0·4509	0·4549	0·4589	0·4630	19	4	8	12	16	20	24	28	32	36
71	·4630	·4671	·4713	·4755	·4797	·4839	·4882	18	4	8	13	17	21	25	29	34	38
72	·4882	·4925	·4969	·5013	·5057	·5102	·5147	17	4	9	13	18	22	26	31	35	40
73	·5147	·5192	·5238	·5284	·5331	·5378	·5425	16	5	9	14	19	23	28	33	37	42
74	·5425	·5473	·5521	·5570	·5619	·5669	·5719	15	5	10	15	20	25	30	34	39	44
75	0·5719	0·5770	0·5822	0·5873	0·5926	0·5979	0·6032	14	5	10	16	21	26	31	37	42	47
76	·6032	·6086	·6141	·6196	·6252	·6309	·6366	13	6	11	17	22	28	33	39	45	50
77	·6366	·6424	·6483	·6542	·6603	·6664	·6725	12	6	12	18	24	30	36	42	48	54
78	·6725	·6788	·6851	·6915	·6980	·7047	·7113	11	6	13	19	26	32	39	45	52	58
79	·7113	·7181	·7250	·7320	·7391	·7464	·7537	10°	7	14	21	28	35	42	49	57	64
80°	0·7537	0·7611	0·7687	0·7764	0·7842	0·7922	0·8003	9	8	16	23	31	39	47	54	62	70
81	·8003	·8085	·8169	·8255	·8342	·8431	·8522	8	9	17	26	35	43	52	61	69	78
82	·8522	·8615	·8709	·8806	·8904	·9005	·9109	7	10	20	29	39	49	59	69	78	88
83	0·9109	0·9214	0·9322	0·9433	0·9547	0·9664	0·9784	6									
84	0·9784	0·9907	1·0034	1·0164	1·0299	1·0437	1·0580	5	Differences vary so rapidly here that tabulation is impossible.								
85	1·0580	1·0728	1·0882	1·1040	1·1205	1·1376	1·1554	4									
86	1·1554	1·1739	1·1933	1·2135	1·2348	1·2571	1·2806	3									
87	1·2806	1·3055	1·3318	1·3599	1·3899	1·4221	1·4569	2									
88	1·4569	1·4947	1·5362	1·5819	1·6331	1·6911	1·7581	1									
89	1·7581	1·8373	1·9342	2·0591	2·2352	2·5363	$+\infty$	0°									
	60′	50′	40′	30′	20′	10′	0′		1′	2′	3′	4′	5′	6′	7′	8′	9′

LOGARITHMS OF COTANGENTS.

ANSWERS

PART ONE

Exercise I

1. 1, 3, 7, 11, 17, 53, 71, 73.

2. (*a*) $2\times2\times2\times2$. (*b*) $2\times2\times2\times2\times2$. (*c*) 71×1. (*d*) 5×17. (*e*) $2\times2\times2\times3\times7$. (*f*) 211×1. (*g*) 251×1. (*h*) $2\times2\times11\times11$. (*i*) 571×1. (*j*) $3\times3\times3\times7\times7\times7$. (*k*) $3\times3\times5\times5\times7\times7$. (*l*) 11×101.

Exercise II

1. (*a*) 12. (*b*) 18. (*c*) 54. (*d*) 21. (*e*) 72. (*f*) 60. (*g*) 24. (*h*) 36. (*i*) 60. (*j*) 36. (*k*) 60. (*l*) 60.

Exercise III

1. (*a*) $\frac{1}{3}$. (*b*) $\frac{1}{2}$. (*c*) $\frac{3}{4}$. (*d*) $\frac{3}{4}$. (*e*) $\frac{2}{3}$. (*f*) $\frac{2}{5}$. (*g*) $\frac{3}{5}$. (*h*) $\frac{25}{28}$. (*i*) $\frac{1}{3}$. (*j*) $\frac{16}{21}$. (*k*) $\frac{4}{7}$. (*l*) $\frac{9}{20}$. (*m*) $\frac{22}{101}$. (*n*) $\frac{2}{3}$. (*o*) $\frac{8}{9}$. (*p*) $\frac{5}{21}$. (*q*) $\frac{13}{23}$. (*r*) $\frac{49}{55}$. (*s*) $\frac{7}{12}$. (*t*) $\frac{7}{11}$. (*u*) $\frac{3}{20}$. (*v*) $\frac{4}{13}$. (*w*) $\frac{31}{47}$. (*x*) $\frac{27}{35}$.

2. (*a*) $\frac{7}{4}$. (*b*) $\frac{7}{2}$. (*c*) $\frac{19}{4}$. (*d*) $\frac{57}{7}$. (*e*) $\frac{34}{5}$. (*f*) $\frac{403}{100}$. (*g*) $\frac{457}{100}$. (*h*) $\frac{117}{22}$. (*i*) $\frac{266}{13}$. (*j*) $\frac{454}{11}$. (*k*) $\frac{197}{15}$. (*l*) $\frac{534}{31}$. (*m*) $\frac{1323}{32}$. (*n*) $\frac{627}{25}$. (*o*) $\frac{107}{18}$. (*p*) $\frac{15}{1}$.

3. (*a*) $1\frac{3}{4}$. (*b*) $2\frac{2}{3}$. (*c*) $3\frac{3}{5}$. (*d*) $15\frac{1}{2}$. (*e*) $1\frac{5}{12}$. (*f*) $3\frac{1}{7}$. (*g*) $5\frac{7}{100}$. (*h*) $10\frac{11}{17}$. (*i*) $27\frac{9}{23}$. (*j*) $10\frac{23}{112}$. (*k*) $38\frac{78}{97}$. (*l*) $39\frac{10}{29}$. (*m*) $58\frac{144}{157}$. (*n*) $94\frac{21}{47}$. (*o*) $17\frac{73}{101}$. (*p*) $83\frac{54}{89}$.

4. (*a*) 20. (*b*) 24. (*c*) 99. (*d*) 51. (*e*) 9, 21. (*f*) 18, 90. (*g*) 21, 9. (*h*) 26, 91, 9.

Exercise IV

1. (*a*) $\frac{3}{4}$. (*b*) $\frac{7}{10}$. (*c*) $\frac{19}{42}$. (*d*) $\frac{17}{30}$. (*e*) $\frac{53}{60}$. (*f*) $\frac{43}{66}$. (*g*) $5\frac{5}{6}$. (*h*) $2\frac{1}{4}$. (*i*) $7\frac{19}{24}$. (*j*) $3\frac{7}{20}$. (*k*) $6\frac{23}{36}$. (*l*) $6\frac{917}{1000}$.

2. (*a*) $\frac{1}{2}$. (*b*) $\frac{4}{21}$. (*c*) $\frac{7}{12}$. (*d*) $\frac{1}{6}$. (*e*) $\frac{7}{36}$. (*f*) $1\frac{1}{2}$. (*g*) $2\frac{1}{8}$. (*h*) $1\frac{7}{8}$. (*i*) $1\frac{19}{24}$. (*j*) $3\frac{9}{10}$. (*k*) $\frac{34}{35}$. (*l*) $\frac{13}{14}$.

3. (*a*) $\frac{7}{20}$. (*b*) $1\frac{91}{1000}$. (*c*) $4\frac{1}{3}$. (*d*) $6\frac{5}{8}$. (*e*) $3\frac{2}{3}$. (*f*) $8\frac{29}{36}$. (*g*) $2\frac{149}{300}$. (*h*) $9\frac{1}{4}$. (*i*) $1\frac{2}{3}$. (*j*) $7\frac{61}{294}$. (*k*) $2\frac{4}{33}$. (*l*) $4\frac{17}{30}$.

4. (*a*) $\frac{11}{42}$. (*b*) $\frac{3}{1000}$. (*c*) $\frac{7}{72}$. (*d*) $\frac{1}{24}$. (*e*) $\frac{1}{84}$. (*f*) $\frac{13}{78}$.

5. $\frac{5}{6}$, $\frac{4}{5}$, $\frac{3}{4}$, $\frac{7}{10}$, $\frac{2}{3}$.

6. (*a*) 375. (*b*) 250. (*c*) 875. (*d*) $312\frac{1}{2}$.

Exercise V

1. 200 ml. 2. 450 ml.

3. (*a*) $8\frac{5}{12}$ ohms. (*b*) $25\frac{23}{30}$ ohms. (*c*) $13\frac{25}{120}$ ohms.

4. (*a*) $\frac{15}{16}$ ohm. (*b*) $1\frac{1}{7}$ ohms. (*c*) $\frac{15}{22}$ ohm. (*d*) $\frac{3}{4}$ ohm.

5. (*a*) 15 mf. (*b*) $1\frac{5}{9}$ mf. (*c*) $\frac{8}{9}$ mf. (*d*) $1\frac{1}{23}$ mf.

6. (*a*) $2\frac{3}{8}$ ohms. (*b*) $2\frac{1}{3}$ mf.

7. $\frac{11}{20}$; 180 gal. 8. (*a*) $\frac{4}{15}$; (*b*) $\frac{2}{15}$ (making $\frac{7}{15}$ in each tank).

Exercise VI

1. (*a*) $\frac{12}{35}$. (*b*) $\frac{9}{40}$. (*c*) $3\frac{3}{4}$. (*d*) 9. (*e*) $\frac{25}{28}$. (*f*) $\frac{27}{55}$. (*g*) $\frac{18}{35}$. (*h*) $\frac{1}{16}$.

2. (*a*) $\frac{5}{6}$. (*b*) $1\frac{1}{2}$. (*c*) 3. (*d*) $11\frac{1}{4}$. (*e*) $6\frac{2}{3}$. (*f*) 63. (*g*) $2\frac{1}{5}$. (*h*) 3.

3. (*a*) $\frac{1}{4}$. (*b*) 9. (*c*) $19\frac{3}{5}$. (*d*) $5\frac{5}{6}$. (*e*) 36. (*f*) $33\frac{3}{4}$.

4. (*a*) 275 ml. (*b*) 169 ml. (*c*) 502 ml. (*d*) 237 ml. (*e*) 372 ml.

Exercise VII

1. (*a*) 5 sec. (*b*) $9\frac{4}{5}$ sec. (*c*) $3\frac{4}{5}$ sec.

2. $18\frac{14}{15}$ m.p.h. 3. 702 lb.

4. (*a*) $4\frac{3}{8}$ sq. ft. (*b*) $18\frac{8}{9}$ sq. in. (*c*) $29\frac{1}{6}$ sq. in. (*d*) $2\frac{29}{48}$ sq. ft. (or 375 sq. in.).

5. (*a*) $\frac{77}{125}$ sq. in. (*b*) $\frac{77}{512}$ sq. in. (*c*) $\frac{99}{224}$ sq. in. (*d*) $\frac{275}{3584}$ sq. in.

6. $\frac{11}{128}$ sq. in.

7. $14\frac{11}{24}$ in. wingspan; $11\frac{1}{24}$ in. length; $3\frac{9}{16}$ in. height. 8. $23\frac{93}{128}$ ft.

9. (*a*) 70° C. (*b*) 85° C. (*c*) $21\frac{2}{3}$° C. (*d*) −10° C. (*e*) 95° F. (*f*) $144\frac{1}{2}$° F. (*g*) $62\frac{3}{5}$° F. (*h*) −13° F.

10. 20.

Exercise VIII

1. (*a*) $\frac{7}{40}$. (*b*) $\frac{3}{20}$. (*c*) 12. (*d*) 40. (*e*) $1\frac{1}{4}$. (*f*) $2\frac{1}{3}$. (*g*) $\frac{30}{49}$. (*h*) $1\frac{8}{17}$.

2. (*a*) $5\frac{1}{3}$. (*b*) 6. (*c*) $5\frac{3}{5}$. (*d*) $1\frac{4}{5}$. (*e*) $\frac{1}{4}$. (*f*) 168.

3. (*a*) 122 m.p.h. (*b*) 152 m.p.h. (*c*) $183\frac{1}{3}$ m.p.h. (*d*) 246 m.p.h.

4. (*a*) $1\frac{3}{4}$ hr. (*b*) $2\frac{1}{2}$ hr. (*c*) $3\frac{1}{6}$ hr. (*d*) $4\frac{1}{10}$ hr.

Exercise IX

1. $3\frac{1}{4}$ in. 2. (*a*) $1\frac{5}{44}$ in. (*b*) $\frac{35}{88}$ in. (*c*) $\frac{63}{88}$ in. (*d*) $\frac{7}{22}$ in.

3. $1\frac{2}{5}$ sec. gaining. 4. $2\frac{1}{5}$ sec. gaining.

5. Midnight June 11/12th. 6. 7 strips; $\frac{1}{8}$ in. left over.

7. 5 pieces. $2\frac{3}{16}$ in. left over.

8. (*a*) 7 hr. (*b*) $8\frac{4}{45}$ hr. (*c*) $14\frac{4}{5}$ hr.

9. (*a*) $38\frac{6}{11}$ gal. per hr. (*b*) $39\frac{13}{23}$ gal. per hr. (*c*) $40\frac{1}{2}$ gal. per hr.

10. $26\frac{8}{17}$ min.

Exercise X

1. (*a*) $\frac{5}{24}$. (*b*) $\frac{3}{40}$. (*c*) $\frac{161}{220}$. (*d*) $\frac{19}{52}$. (*e*) $\frac{25}{102}$. (*f*) $\frac{1}{63,360}$.

2. $\frac{1}{316,800}$. 3. (*a*) $\frac{1}{63,360}$. (*b*) $\frac{1}{10,560}$. (*c*) $\frac{1}{253,440}$.

4. (*a*) $\frac{1}{3}$ ml. to 1 in. (*b*) $2\frac{1}{2}$ ml. to 1 in. (*c*) $1\frac{2}{5}$ ml. to 1 in.

5. $\frac{29}{64}$. 6. $\frac{95}{108}$. 7. (*a*) $\frac{1}{3}$. (*b*) 3 hr.

8. $\frac{33}{38}$. 9. $\frac{50}{127}$. 10. 275 m.p.h.

11. $\frac{1}{5}$. 12. 972 lb.

Exercise XI

1. (*a*) 21·3328. (*b*) 55·183. (*c*) 85·6191. (*d*) 97·22125. (*e*) 285·233. (*f*) 1·7512. (*g*) 379·9707. (*h*) 405·071. (*i*) 80·767. (*j*) 276·21335.

2. (*a*) 6·19. (*b*) 4·449. (*c*) 5·208. (*d*) 5·124. (*e*) 1·823. (*f*) 0·2246. (*g*) 6·19. (*h*) 88·898. (*i*) 36·423. (*j*) 0·090911. (*k*) 0·00101.

3. 1·19 in. 4. 2·59 in.

5. (*a*) 1045·8 mb. (*b*) 931 mb. 6. (*a*) 40·08 in. (*b*) 2·98 in.

Exercise XII

1. (*a*) 5. (*b*) 4. (*c*) 7. (*d*) 2. (*e*) 6. (*f*) 7.

2. (*a*) 5·1. (*b*) 403. (*c*) 88·2. (*d*) 42·4. (*e*) 7·502. (*f*) 0·1458. (*g*) 35·16. (*h*) 0·5412.

3. (*a*) 5·936. (*b*) 0·2782. (*c*) 475·64. (*d*) 0·0117. (*e*) 197·892. (*f*) 0·78694. (*g*) 9·984. (*h*) 2684·834. (*i*) 30·72876. (*j*) 4·4457. (*k*) 0·3621528. (*l*) 54·74794. (*m*) 0·0214731. (*n*) 17·637529.

4. (*a*) 6·89 sq. in. (*b*) 13·8125 sq. in. (*c*) 23·4241 sq. in.

5. (*a*) 29·2 in. (*b*) 28·8 in. (*c*) 30·1 in.

6. (*a*) 7·08° C. (*b*) 5·1° C. (*c*) −4·8° C. (*d*) −56·28° C.

7. (*a*) 813·2 mb. (*b*) 773·2 mb. (*c*) 613·2 mb. (*d*) 833·2 mb. (*e*) 903·2 mb. (*f*) 948·2 mb.

8. (*a*) 6990 ft. (*b*) 10,896 ft. (*c*) 2604 ft. (*d*) 3396 ft. (*e*) 4530 ft.

Exercise XIII

1. (*a*) 21. (*b*) 31·2. (*c*) 1·2. (*d*) 13. (*e*) 263·1. (*f*) 0·00248. (*g*) 2020. (*h*) 5·4. (*i*) 81·7. (*j*) 7·8. (*k*) 4900. (*l*) 3764. (*m*) 0·24. (*n*) 0·04. (*o*) 0·00525. (*p*) 10,300.

2. (*a*) 0·081. (*b*) 0·358. (*c*) 0·20. (*d*) 0·747. (*e*) 0·00. (*f*) 0·00060. (*g*) 0·010. (*h*) 0·0011. (*i*) 4. (*j*) 62.

3. 4·81 in. 4. 6879 N.M. 5. 14·0 in.

6. (*a*) 0·985. (*b*) 0·866. (*c*) 0·644. (*d*) 0·343.

7. 1·44 ohms. 8. 0·568 ohm.

9. 152 pieces; 5·6 ft. left over. 10. 17 pieces; 0·31 in. left over.

Exercise XIV

1. (*a*) $\frac{1}{2}$. (*b*) $\frac{19}{20}$. (*c*) $\frac{3}{20}$. (*d*) $\frac{47}{50}$. (*e*) $\frac{7}{8}$. (*f*) $\frac{3}{40}$. (*g*) $1\frac{7}{25}$. (*h*) $5\frac{13}{20}$. (*i*) $2\frac{9}{20}$. (*j*) $1\frac{1}{40}$. (*k*) $2\frac{1}{80}$. (*l*) $10\frac{1}{1000}$. (*m*) $3\frac{63}{125}$. (*n*) $10\frac{3}{8}$. (*o*) $\frac{16}{625}$. (*p*) $3\frac{71}{500}$. (*q*) $6\frac{1}{16}$. (*r*) $2\frac{11}{250}$. (*s*) $1\frac{51}{125}$. (*t*) $\frac{11}{10000}$.

2. (*a*) 0·857. (*b*) 0·625. (*c*) 0·312. (*d*) 0·461. (*e*) 0·944. (*f*) 0·718.

3. 0·366. 4. 0·813. 5. 0·914.

6. 0·868. 7. 0·555. 8. 0·029.

Exercise XV

1. $1\frac{31}{60}$. 2. $1\frac{13}{60}$. 3. $1\frac{41}{120}$. 4. $1\frac{4}{15}$. 5. $\frac{2}{3}$.

6. $\frac{3}{8}$. 7. $2\frac{2}{3}$. 8. $\frac{43}{53}$. 9. $\frac{11}{16}$. 10. $1\frac{11}{21}$.

Exercise XVI

1. 21·6. 2. 0·185. 3. 0·138. 4. 1·033. 5. 4·25.

6. 0·916. 7. 6·12. 8. 0·025. 9. 0·814. 10. 4·166.

Exercise XVII

1. $\frac{1}{2}$. 2. 8 min. past 10 a.m. 3. After $16\frac{2}{3}$ min.

4. (*a*) 14 min. 15 sec. after noon. (*b*) $28\frac{1}{2}$ miles.

5. (*a*) 3.48 p.m. (*b*) 33 ml. and 21 ml. respectively.

6. 150 m.p.h. and 90 m.p.h. 7. 81 m.p.h.

8. (*a*) 2.56 p.m. (*b*) 200 closing. 9. (*a*) 27·5 knots. (*b*) 10.20 a.m.

10. (*a*) 11.30 a.m. (*b*) 11.50 a.m. (*c*) $83\frac{1}{3}$ N.M. 11. 1 ml. 12. 1 ml.

Exercise XVIII

1. (*a*) $\frac{1}{1000}$, 0·001. (*b*) $\frac{1}{1000}$, 0·001. (*c*) $\frac{1}{100}$, 0·01. (*d*) $\frac{3}{10}$, 0·3. (*e*) $\frac{1}{250}$, 0·004. (*f*) $\frac{1}{2500}$, 0·0004.
2. (*a*) 5·004. (*b*) 1·156. (*c*) 0·056. (*d*) 0·75. (*e*) 0·0024. (*f*) 0·00446.
3. (*a*) 246,200. (*b*) 2·613. (*c*) 4·28. (*d*) 22,000. (*e*) 0·76. (*f*) 0·0466. (*g*) 3,002,000. (*h*) 0·06675. (*i*) 0·000007. (*j*) 2400.
4. (*a*) 8·27 m. (*b*) 420·724 m. (*c*) 197,561 mm. (*d*) 0·9822 Km. (*e*) 2721 gm. (*f*) 4,016,500 c.c.

Exercise XIX

1. (*a*) 101·4 Km. (*b*) 344·4 Km. (*c*) 130·7 Km. (*d*) 160·9 Km.
2. (*a*) 62·1 ml. (*b*) 353·1 ml. (*c*) 20·1 ml. (*d*) 111·1 ml.
3. 9146 m.
4. (*a*) Wellington by 2054 ft. (*b*) Breda by 3021 ft.
5. *A* by 0·92 ft.
6. 1·851 Km.
7. 0·540.
8. 43,965 ft.
9. 0·394.
10. 0·621.
11. 10·75 sq. ft.
12. 0·093.

Exercise XX

1. (*a*) 220. (*b*) 132. (*c*) 99. (*d*) 550.
2. (*a*) 4546. (*b*) 2727·6. (*c*) 2045·7. (*d*) 19·7.
3. 166$\frac{2}{3}$ min.
4. *A* by 2·84 gal. per hour.
5. 273 gal.
6. $\frac{273}{300}$.
7. Plane *B* by 5 gal.
8. Pump *A* by 0·28 gal. per min.
9. 0·055.
10. 220.

Exercise XXI

1. 78·9 Kgm.
2. 720 Kgm.
3. *B* by 57·7 lb.
4. Lifting surface 55·76 sq. metres; weight 419 Kgm.; wing loading 7·53 Kgm. per sq. metre.
5. (*a*) 41·3 lb. per sq. ft. (*b*) 42·3 lb. per sq. ft.
6. Rope circumference = 1·905 cm. Breaking load = 1·72 metric tonnes.

Exercise XXII

1. (*a*) 246 S.M. (*b*) 748 S.M. (*c*) 18 S.M. (*d*) 2317 S.M. (*e*) 664 S.M.
2. (*a*) 177 N.M. (*b*) 543 N.M. (*c*) 15 N.M. (*d*) 1728 N.M. (*e*) 489 N.M.

Exercise XXIII

1. 335 N.M. = 5° 35′ N. Lat. arrival 26° 5′ N.
2. 238 N.M. = 3° 58′ S. „ „ 2° 43′ S.
3. 542 N.M. = 9° 2′ N. „ „ 5° 5′ N.
4. 363 N.M. = 6° 3′ N. „ „ 58° 27′ N.
5. 1169 N.M. = 19° 29′ S. „ „ 31° 2′ N.
6. 65 N.M. = 1° 5′ N. „ „ 38° 54′ S.

Exercise XXIV

1. Diff. in lat. 2° 33′ S. = 153 N.M. = 176 S.M.
2. „ „ 5° 12′ N. = 312 N.M. = 359 S.M.
3. „ „ 3° 12′ N. = 192 N.M. = 221 S.M.
4. „ „ 5° 22′ S. = 322 N.M. = 371 S.M.
5. „ „ 10° 4′ N. = 604 N.M. = 696 S.M.
6. „ „ 3° 58′ S. = 238 N.M. = 274 S.M.

Exercise XXV

1. (*a*) 283·97 ml. per day. (*b*) 171·8 ml. per day.
2. 913·2 ml.
3. 256·8 ml.
4. 237 m.p.h.
5. 189·6 m.p.h.
6. 95·9 m.p.h.
7. 232 m.p.h.
8. 2460 ft.
9. 6 ft. $8\frac{13}{16}$ in.
10. 31° 15′ 20″.
11. 29·17 in. mercury.
12. 1375 ft.
13. 0·205 mb. per mile falling.
14. 1003·4 mb.
15. 0·094 mb. per mile rising.

Exercise XXVI

1. 18 sec.
2. Zero.
3. −9·6° C.
4. 10·4° F.
5. 16° 21·5′ N.
6. 2° 31′ N.
7. 656 acres.
8. 1 p.m.
9. 1 ton 15 cwt. 3 qr.
10. 2.25 a.m.
11. 5.12 p.m.
12. (*a*) by $2\frac{2}{17}$ kt.

Exercise XXVII

1. (*a*) Cr. sp. 165 m.p.h. (*b*) Max. sp. 205·7 m.p.h. (*c*) Max. sp. 266·6 m.p.h. (*d*) Cr. sp. 218·4 m.p.h.
2. (*a*) $\frac{9}{10}$. (*b*) $\frac{7}{8}$. (*c*) $\frac{22}{25}$. (*d*) 38·8 gal. (*e*) 100 gal.
3. (*a*) $1\frac{83}{359}$, 1·24. (*b*) $1\frac{6}{53}$, 1·113. (*c*) $1\frac{25}{93}$, 1·267. (*d*) $1\frac{11}{116}$, 1·066.

4. (*a*) 1·54 lb. per sq. ft. (*b*) 19·5 lb. per sq. ft. (*c*) 36 lb. per sq. ft. (*d*) 31·7 lb. per sq. ft. (*e*) 20·7 lb. per sq. ft. (*f*) 32 lb. per sq. ft.

5. (*a*) 14,398 lb. (*b*) 516 sq. ft.

6. (*a*) 31·4 in. (*b*) 10·99 in. (*c*) 13·502 cm.

7. 245 miles.

8. (*a*) 7·64 in. (*b*) 2·07 cm. (*c*) 10·98 in.

9. 53 ml.

10. 28·26 ml.

11. (*a*) 1·64 lb. per H.P. (*b*) 1·75 lb. per H.P. (*c*) 2·03 lb. per H.P. (*d*) 2·09 lb. per H.P.

12. (*a*) 603 H.P. (*b*) 605 H.P.

13. (*a*) 1060 lb. (*b*) 366 lb.

14. (*a*) Speed ratio $\frac{3}{2}$, 450 r.p.m. (*b*) $\frac{7}{3}$, 560 r.p.m. (*c*) $\frac{8}{3}$, 112·5 r.p.m. (*d*) $\frac{2}{3}$, 360 r.p.m. (*e*) $\frac{3}{2}$, 12 cogs. (*f*) $\frac{7}{8}$, 32 cogs.

15. (*a*) 330 yd., 550 yd., 880 yd. (*b*) 160 yd., 560 yd., 1040 yd. (*c*) $547\frac{5}{9}$ yd., $586\frac{2}{3}$ yd., $625\frac{7}{9}$ yd.

16. 100 miles.

17. 5 : 1; 1 : 5.

18. (*a*) 1 : 1; 5 amp. each. (*b*) 10 : 1; $9\frac{1}{11} : \frac{10}{11}$ amp. (*c*) 100 : 1; $9\frac{91}{101} : \frac{10}{101}$ amp.

19. (*a*) (i) 400 : 1; (ii) 6000 volts. (*b*) (i) 400 : 1; (ii) 4000 volts. (*c*) (i) 100 : 3; (ii) 400 volts.

20. (*a*) 50 : 1. (*b*) 1600 turns.

21. (*a*) 3000. (*b*) 3000. (*c*) 6000. (*d*) 8000. (*e*) 4000. (*f*) 5000. (*g*) 1000 volts. (*h*) 125 : 1.

Exercise XXVIII

1. (*a*) 3·6 ohms. (*b*) 2·09 ohms. (*c*) 6·68 ohms. (*d*) 1·88 ohms.

2. (*a*) 560·3 yd. (*b*) 8·98 yd. (*c*) 1·2244 ml. (*d*) 204·7 yd.

3. (*a*) 226·3 m.p.h. (*b*) 90 m.p.h. (*c*) 100 m.p.h. (*d*) 384 m.p.h.

4. (*a*) 140 ml. (*b*) 253·1 ml. (*c*) 14 ml. (*d*) 158·6 N.M. = 182·6 ml. (*e*) 1159·2 Km. = 720·1 ml.

5. 833·3 yd.

6. 66 gal.

7. 525 ml.

8. (*a*) 2 hr. 18 min. (*b*) 2 hr. 2 min. (*c*) 2 hr. 2 min. (*d*) 1 hr. 4 min.

Exercise XXIX

1. 1·68 ohms.

2. 0·87 ohm.

3. 12 days.

4. 144 tons.

5. 5·6 amp.

6. 3·5 amp.

7. 1·1 amp.

8. None.

Exercise XXX

1. (*a*) 5 ml. (*b*) 11·33 ml. (*c*) 17·34 ml. (*d*) 9·33 ml.
2. (*a*) 12 ml. (*b*) 22 ml. (*c*) 28 ml. (*d*) 36·67 ml.
3. (*a*) 23·47 ml. (*b*) 14·93 ml. (*c*) 12·25 ml. (*d*) 19·25 ml. (*e*) 38·4 ml. (*f*) 65 ml. (*g*) 62·9 ml.

Exercise XXXI

1. (*a*) $\frac{3}{20}$. (*b*) $\frac{18}{25}$. (*c*) $\frac{1}{16}$. (*d*) $1\frac{1}{40}$. (*e*) $\frac{3}{80}$. (*f*) 1.
2. (*a*) 0·16. (*b*) 0·025. (*c*) 1·25. (*d*) 0·0375. (*e*) 0·142. (*f*) 0·0105.
3. (*a*) 18 %. (*b*) 210 %. (*c*) 34 %. (*d*) 70 %. (*e*) 6·5 %. (*f*) 8·4 %.
4. (*a*) 50 %. (*b*) 25 %. (*c*) 75 %. (*d*) $12\frac{1}{2}$ %. (*e*) $87\frac{1}{2}$ %.
5. (*a*) 33·3 %. (*b*) 83·3 %. (*c*) 22·2 %. (*d*) 23·0 %. (*e*) 7·8 %. (*f*) 65·9 %.
6. (*a*) 198 m.p.h. (*b*) 163·4 m.p.h. (*c*) 99·36 knots. (*d*) 248·04 Km. per hour. (*e*) 185·12 knots.
7. (*a*) 12·5 % decrease. (*b*) 13·7 % increase. (*c*) 12·7 % decrease. (*d*) 17·2 % decrease.
8. (*a*) $32\frac{1}{2}$ %. (*b*) 15·1 %.
9. (*a*) (i) 87 %; (ii) 21·6 %. (*b*) (i) 83 %; (ii) 28·8 %. (*c*) (i) 83 %; (ii) 22·4 %.
10. 30·7 %.
11. (*a*) 130·5 m.p.h. (*b*) 141 m.p.h. (*c*) 149·3 m.p.h. (*d*) 175·64 m.p.h. (*e*) 110·8 m.p.h. (*f*) 113·3 m.p.h.
12. Range increased by 2·8 %.
13. 13·1 % too low.
14. 15·1% too high.
15. (*a*) 57·68 yd. (*b*) 192·61 yd. (*c*) 906·4 yd. (*d*) 81·37 Km.
16. Aluminium, 94·7 %; Copper, 4·2 %; Manganese, 0·6 %; Magnesium, 0·5 %.
17. Aluminium, 92·6 %; Copper, 3·9 %; Nickel, 2·1 %; Magnesium, 1·4 %.
18. 3·03 % high.
19. 15,300 ft.
20. 3·006 in.

Exercise XXXII

1. (*a*) $5b$. (*b*) $-2x$. (*c*) $12y$. (*d*) $3z$. (*e*) $-6p$.
2. (*a*) 6. (*b*) $-1\frac{1}{2}$. (*c*) 0.
3. (*a*) $5s-t+4$. (*b*) $9m+4n+6$. (*c*) $5w-12v+5$.
4. (*a*) 10. (*b*) $5\frac{1}{2}$. (*c*) 6. (*d*) 7. (*e*) 14.
5. (*a*) 24. (*b*) 12. (*c*) $-2\frac{1}{2}$. (*d*) 0. (*e*) 3. (*f*) 2. (*g*) $\frac{5}{12}$. (*h*) 0·5.

6. (a) $3a \times b \times b \times c \times c \times c$. (b) $5x \times x \times x \times y \times z \times z$. (c) $2a \times a \times a \times b \times c$. (d) $5a \times p \times p \times q \times q$.

7. (a) 4. (b) -9. (c) 72. (d) $2\frac{1}{2}$.

8. (a) $12x^4y^3s^2t$. (b) $8a^3b^3c^3d$. (c) $3m^4n^2xy$. (d) $a^3c^2q^4r$.

9. (a) $4b^2x^4r^3$. (b) $6c^4d^3e^5$. (c) $s^2t^2v^6$. (d) $a^4b^4c^2$.

Exercise XXXIII

1. (a) $4a-4c$. (b) $9x-7y+4z$. (c) $6m^2+9an+2p^2+5p$. (d) $2x^2z+6y^2z$. (e) $4a^3y+8a^2y^2+2ay^3$. (f) $3p^2q+5p^3+3pq^2$. (g) $\frac{5p}{4}$. (h) $3{\cdot}1ax^2+1{\cdot}4ax-0{\cdot}9a^2x$.

2. (a) $7b+c+d$. (b) $a^2-2ab+5b^2$. (c) $8x^2y-8xy^2$. (d) $4p^2qr+2pq^2r$. (e) x^4-3x^2+3x+5. (f) $2ab$. (g) $3a^2-2ab-b^2$. (h) $ac-ab-bc-1$.

3. (a) abc^2. (b) pqr^2. (c) $2ad^2$. (d) $\frac{a^2r^2}{2}$. (e) ar. (f) $\frac{3\pi r^2x^2}{2}$. (g) a^2r^2. (h) ar^3.

4. (a) $3a^2b^2c^2$. (b) $4a^3b^3c$. (c) $2x^2y^3z$. (d) πa^2p^2. (e) $\frac{3ax^2}{2p}$. (f) $2xyz^3$. (g) $\frac{2a}{p^2}$. (h) $\frac{1}{c^2}$.

5. (a) $3x+3y$. (b) $8x-4a$. (c) $3x+2y$. (d) $3a+3b+3c$. (e) $-4x+4y$. (f) $-5a-10b+\frac{5c}{2}$. (g) $3ax+3by$. (h) $3a^2x+3aby$.

6. (a) $2(x+y)$. (b) $3(a+b+2c)$. (c) $4(x^2+y^2)$. (d) $3x(b-a)$. (e) $a(x+y+z)$. (f) $b(x^2+2xy+y^2)$. (g) $3p(x^2+2y^2)$. (h) $-3ar(a+2r)$.

7. (a) x^2-y^2. (b) $a^2+2ab+b^2$. (c) $p^2+2pq+q^2$. (d) a^4-x^4. (e) $a^3-ax^2+a^2x-x^3$. (f) $2p^2+5pq-pr+2q^2+qr-r^2$. (g) $9a^2+18ax+9x^2$. (h) $\frac{a^2}{8}-\frac{x^2}{8}$.

8. (a) $(a+c)(b+c)$. (b) $(x+y)(y-z)$. (c) $(p-2q)(2q+r)$. (d) $(a^2+b^2)(3p+q)$. (e) $(a+b)(x+y)$. (f) $(c+d)(x+y)$. (g) $(x+y)(a+x)$. (h) $(p+q)(p+q)$.

Exercise XXXIV

1. (a) $r=\frac{C}{2\pi}$. (b) $f=12C^2$. (c) $l=\frac{8t^2}{\pi^2}$. (d) $n=\frac{5H}{2d^2}$. (e) $f=\frac{s-u}{t}$. (f) $f=\frac{2}{t^2}(s-ut)$.

2. (a) $a=\frac{1}{t}\left(\frac{V}{v}-1\right)$. (b) $p=\frac{100(S-c)}{c}$. (c) $h=\frac{S}{2\pi r}-r$.
(d) $a=\frac{S}{n}-\frac{d}{2}(n-1)$. (e) $u=v-\frac{2gS}{W}$. (f) $l=\frac{3h}{2\pi}(Pg-2\pi h)$.
(g) $d=\frac{AKN}{36\times 10^5 \pi s}$. (h) $L=\frac{1}{4\pi^2 f^2 C}$.

3. (a) $v=\frac{ru}{2u-r}$. (b) $n=\frac{CR}{E-rC}$. (c) $x=\frac{dm-bn}{an-cm}$.
(d) $x=\frac{2by}{a}\left(\frac{1+c}{1-c}\right)$. (e) $\mu=\frac{a(R_1-R_2)}{d(R_1+R_2)}$. (f) $t=\frac{(p+q)(x+y)}{2y}$.

4. (a) $P=\frac{RT}{V-b}-\frac{a}{V^2}$. (b) $r=\frac{h}{3}+\frac{V}{\pi h^2}$. (c) $m=\frac{M}{6}\frac{(2V^2-3v^2)}{(v^2-V^2)}$.
(d) $l=\frac{S-\pi(pg+f)}{\pi g}$. (e) $l=\frac{k-R}{2\pi}$. (f) $a=\frac{bp^2}{b-p^2}$.
(g) $M=\frac{8\pi^2 a^2 W}{b(gt^2-4\pi^2 b)}$.

5. (a) $R=\frac{t}{2}+\frac{V}{2\pi lt}$; $1\frac{5}{44}$. (b) $x=\frac{p}{mt-pt-ms}$; $\frac{1}{10}$.
(c) $h=\frac{2Vr}{Sr-2V}$. (d) $m=\frac{a}{b}$. (e) $v=\frac{s}{t}+\frac{1}{2}ft$, (f) $a=\frac{b^2-4}{6}$.

Exercise XXXV

5. (a) E. (b) K. (c) F. (d) L. (e) G. (f) O.
6. (a) 1·5; 3·4. (b) 1·65; 2·1. (c) 1·175; 4·2. (d) 2·325; 3·45.
(e) 1·25; 1·5. (f) 2·0; 3·0.
7. (a) ABWX 1933. (b) ABWX 2805. (c) ABUV 2852.
(d) EFUV 0225. (e) CDWX 5620. (f) CDWX 2042.
8. (a) Q. (b) OA. (c) NT. (d) O. (e) S. (f) L.

Exercise XXXVI

1. (a) 1·7. (b) 0·8. (c) 2·9. (d) 1·9. (e) 3·75. (f) 2·75.
2. (a) 50. (b) 550. (c) 330. (d) 310. (e) 140. (f) 417.
3. (a) 32·2 Km. (b) 34·8 ml. 4. (a) 16·5 N.M. (b) 21·9 ml.
5. (a) 45° C. (b) 140° F. 6. (a) 8·1° C. (b) 2525 ft.
8. (a) 20 min. (b) 28 min. (c) 43 min. (d) 13 min. (e) 70 ml.
(f) 20 ml. (g) 38 ml. (h) 92 ml.
9. (a) 1·07 p.m. (b) $12\frac{1}{3}$ ml.
10. (a) 5·5 N.M. (b) 7·1 N.M. (c) 28 ft. (d) 12 ft.

11. (*a*) 15 min. (*b*) 27 min. (*c*) 8 min. (*d*) 740 ft. p.m. (*e*) 490 ft. p.m. (*f*) 120 ft. p.m. (*g*) 600 ft. p.m. (*h*) 400 ft. p.m. (*i*) 200 ft. p.m.
12. (*a*) 35·75 sq. in. (*b*) 29·75 sq. in. (*c*) Base = Height.
13. 238 m.p.h. 240°, 40 m.p.h. 14. 8·9 amp. 15. 3° E.

Exercise XXXVII

1.

	$\angle ABC$	$\angle BCA$	$\angle CAB$	Sum
(*a*)	36½°	95½°	48°	180°
(*b*)	12°	62°	106°	180°
(*c*)	74°	86°	20°	180°

2. 28·3 ft.
3. 26 fathoms.
4. 33½°.
5. 18 ft.
6. 6·4 ft.
7. $AC = 2{\cdot}06$ in.: $BC \doteq 2{\cdot}98$ in.
8. 111 m.p.h.; 117° T.
9. 26·2 ml.
10. (*a*) Red 116½°; 3·3 ml. (*b*) Red 124°; 5·4 ml.

Exercise XXXVIII

1. $BC = 21\frac{1}{2}$, $CD = 29$, $DE = 6$, $EF = 20$, $FG = 19\frac{1}{2}$, $GH = 15\frac{1}{2}$, $HK = 23$, $KB = 37\frac{1}{2}$.
2. 21½ ml.

Exercise XXXIX

1. (*a*) 51° 33′ N., 1° 16′ E. (*b*) 52° 56′ N., 1° 20′ E. (*c*) 52° 37′ N., 1° 57′ E. (*d*) 52° 12′ N., 0° 8′ E. (*e*) 51° 1′ N., 1° 54′ E.
2. 9¾° W.
3. (*a*) 36 ml. (*b*) 41 ml. (*c*) 75 ml. (*d*) 75 ml. (*e*) 107 ml.
4. (*a*) 31½ N.M. (*b*) 35 N.M. (*c*) 61½ N.M. (*d*) 65 N.M. (*e*) 90 N.M.
5. (*a*) 123°. (*b*) 303°. (*c*) 278°. (*d*) 124°. (*e*) 335°.

Exercise XL

1. 2½ ml.
2. 21½ ml.
3. 6·2 ml.
4. 4·35 ml.
5. 71 ml.
6. 27·6 S.M.
7. 144 sq. ml.
8. 4·06 in.
9. 1·52 in.
10. 3·78 in.

Exercise XLI

1. 10·2 knots, 078½° T.
2. 10·1 knots, 237° T.
3. 89·5 m.p.h., 095° T.
4. (*a*) 136 m.p.h., 179° T. (*b*) 109 m.p.h., 295° T. (*c*) 54 m.p.h., 276° T. (*d*) 96 m.p.h., 216° T. (*e*) 61 m.p.h., 359° T. (*f*) 128 knots, 074° T. (*g*) 89 knots, 141° T.

Exercise XLII

1. 32 m.p.h., 245°.
2. 30 m.p.h., 111°.
3. 41 m.p.h., 164°.
4. 38½ m.p.h., 105°.
5. 44 knots, 177°.
6. 24 m.p.h., 352°.
7. 43 knots, 198°.
8. 32 m.p.h., 243°.
9. 39 m.p.h., 171°.
10. 48 m.p.h., 267°.

Exercise XLIII

1. 314° T, 112 m.p.h.
2. 107° T, 115 m.p.h.
3. 049° T, 127 m.p.h.
4. 237½° T, 112 m.p.h.
5. 130° T, 115 knots.
6. 257½° T, 117 m.p.h.
7. 299° T, 113 m.p.h.
8. 333½° T, 106 knots.
9. 078° T, 116½ m.p.h.
10. 086° T, 102 knots.

Exercise XLIV

1. 3 ml. to 1 in.
2. 6 ml. to 1 in.
3. 2 in. to 1 ml.
4. 3 ml. to 1 in.

Exercise XLV

1. (*a*) 4·242 in. (*b*) 7·07 in. (*c*) 15·554 cm.
2. (*a*) 18·382 in. (*b*) 2·4038 in. (*c*) 3·162 in. (*d*) 37.414 ft. (*e*) 10·605 in.
3. (*a*) 4. (*b*) 9. (*c*) 15. (*d*) 15. (*e*) 10.
4. 62 ft. 6 in.
5. 10 ft.
6. 136 ft. 6 in.
7. 1560 ft.
8. 2750 ft.
9. 6 min.
10. 27 miles.
11. 60 miles.
12. 105 ml.
13. 260 ft.
14. 36·75 ml.
15. 2500 ft.
16. 3000 ft.

PART TWO

Exercise XLVI

1. (*a*) 1·745^c. (*b*) 0·4886^c. (*c*) 0·08725^c. (*d*) 0·061075^c. (*e*) 0·02821^c. (*f*) 0·047115^c. (*g*) 0·05456^c.
2. (*a*) 63° 1′ 31″. (*b*) 145° 11′ 15″. (*c*) 2·06″. (*d*) 10·31″. (*e*) 1° 20′ 40·9″.
3. 0·489 in.
4. 0·157 in.
5. 12·04 in.
6. 0·924 ml., or 1626 yd.
7. 2·6 ml.
8. 318·3 yd.
9. 1° 54′.
10. 6·98 ft.
11. 27′ 29″.

Exercise XLVII

1. 160° C.
2. 180 ml.
3. 320 gal. and 180 gal.
4. 500 gal. and 360 gal.
5. 60 ft. and 30 ft.
6. 10½ in.
7. 150 m.p.h.
8. 140 m.p.h.
9. 0·10005 ohm.
10. 1⅜ mf.

Exercise XLVIII

1. A = 360 m.p.h., B = 180 m.p.h.
2. Air speed = 100 m.p.h., wind speed = 20 m.p.h.
3. 480 gal. and 320 gal.
4. 80 yd. and 40 yd.
5. 5 and 8.
6. 2080 ft. and 2040 ft.
7. 1010 mb. and 15,150 ft.
8. 8 cm. and 3 cm.
9. 6 ohms and 3 ohms.
10. Leak resistance 250 ohms. Resistance of A to leak = 50 ohms.

Exercise XLIX

1. 4 or −6.
2. 160 m.p.h. or 20 m.p.h.
3. 180 m.p.h.
4. 5 cm. and 7 cm.
5. 36 sq. cm.
6. 8.
7. 1½ or −⅔.
8. 10 yd. and 25 yd.
9. 30 m.p.h.
10. 30 m.p.h.

Exercise L

1. $W=\dfrac{At}{t+2x}$.
2. When $t=4$ min., $W=33$ m.p.h., 37 m.p.h., 40 m.p.h. When $t=6$ min., $W=43$ m.p.h., 47 m.p.h., 51 m.p.h.
3. (*a*) $D=VT$. (*b*) $D=\dfrac{V}{12}$. (*c*) $D=\dfrac{Vx}{60}$.

4. $P=\left(\frac{100-x}{100}\right)G.$ 5. $V=\frac{60x}{y}.$ 6. $T=\frac{60x}{M}.$

7. $D=z-\frac{xy}{60}.$ 8. $G=\frac{2xPT}{7\cdot 2}.$ 9. $A=\frac{xy+ab}{x+a}.$

10. $(y-x)$ hours $=\frac{ay-bx}{c}.$ 11. $E=\frac{T(100-y)}{100x}.$

12. (a) $T=D\left(\frac{V_1+V_2}{V_1V_2}\right).$ (b $T=\frac{G}{y}.$ (c) $D=\frac{V_1V_2G}{(V_1+V_2)\,y}.$

13. (a) $R=2\sqrt{y^2-x^2}.$ (b) $R=\sqrt{4y^2-\left(\frac{12x+z}{6}\right)^2}.$

14. $d=\sqrt{h^2+2hR}.$

Exercise LI

1. 11·4 knots, 295° T; ahead 2·18 ml.
2. 14·2 knots, 091° T.
3. 23·6 knots, 196½° T; astern 5·4 ml.
4. 10 knots, 079° T.
5. 9·4 knots, 152° T.
6. (a) 158° T. (b) 7·3 cables. (c) 1·9 min.
7. (a) 059° T. (b) 7·3 cables. (c) 5·5 min. (nearly).
8. (a) 67 m.p.h. (b) 062° T. (c) 18 min. (d) 121 m.p.h. (e) 319° T. (f) 10 min.
9. (a) 019° T. (b) 14·5 m.p.h. (c) 37 min. (d) 260° T. (e) 147 m.p.h. (f) 3·7 min.
10. (a) 1202. (b) 1202·6 (c) 1203·2. (d) 1·3 ml.
11. 1 hr. 5½ min.
12. (a) 098° T. (b) 5·5 knots. (c) 6½ min.

Exercise LV

1. 3954. 2. 2020. 3. 3·221. 4. 0·7675.
5. 2·501. 6. 0·00000084. 7. 0·1167. 8. 1·103.
9. 16·62. 10. 45·93.

Exercise LVI

1. 6·337. 2. 151·2. 3. 0·0003334. 4. 0·7403.
5. 0·0001123. 6. 0·0004482. 7. 0·002273. 8. 100.
9. 0·01. 10. 0·509.

Exercise LVII

1. 1·735. 2. 1·412. 3. 3068 metres. 4. 110·6.
5. 87·44 N.M. 6. 0·000277. 7. 161·5 ml. 8. 525·4 Km.
9. 105·9 m.p.h. 10. 1 hr. 42 min.

Exercise LVIII

1. 10·48. 2. 4·385. 3. 56·72. 4. 10·36.
5. 13·5. 6. 5·994. 7. 0·0785. 8. 3·718.
9. 2·818. 10. 9·452.

Exercise LIX

1. 9·078 sq. in. 2. 67·4 sq. in. 3. 21·53 cu. ft.
4. 74·01 cu. in. 5. 11·23 cu. n. 6. 102·7 sq. in.
7. 2·042 in. 8. 2·997 in. 9. 7·691 sq. in.
10. 7·528 sq. in.

Exercise LX

1. (*a*) 56° 50′. (*b*) 42° 35′. (*c*) 27° 13′. (*d*) 16° 39′.
2. (*a*) 0·1794, 79° 40′. (*b*) 0·9805, 11° 20′. (*c*) 0·2987, 72° 37′. (*d*) 0·5415, 57° 13′. (*e*) 0·5237, 58° 25′. (*f*) 0·9260, 22° 11′.
3. (*a*) 0·9703. (*b*) 0·9596. (*c*) 0·3448. (*d*) 0·8445. (*e*) 0·7876. (*f*) 0·6530.
4. (*a*) $\bar{1}$·5767. (*b*) $\bar{1}$·6716. (*c*) $\bar{1}$·7503. (*d*) $\bar{1}$·9875. (*e*) $\bar{1}$·6321. (*f*) $\bar{1}$·9537. (*g*) $\bar{1}$·9404. (*h*) $\bar{1}$·9329.
5. (*a*) 0·6009. (*b*) 0·9110. (*c*) 1·5439. (*d*) 2·532. (*e*) 1·1504. (*f*) 1·0477. (*g*) 0·7785. (*h*) 1·2066.
6. (*a*) $\bar{1}$·8125. (*b*) $\bar{1}$·9570. (*c*) 0·0934. (*d*) $\bar{1}$·8552. (*e*) 1·7581. (*f*) 1·0034. (*g*) 0·5416. (*h*) 0·1126.
7. (*a*) 0·5191. (*b*) 0·9018. (*c*) 4·81. (*d*) 1·325. (*e*) 1·939. (*f*) 1·469. (*g*) 1·638. (*h*) 1·996.
8. (*a*) 146°. (*b*) 137° 18′. (*c*) 88° 41′. (*d*) 122° 3′.
9. (*a*) +. (*b*) −. (*c*) −. (*d*) −. (*e*) +. (*f*) +. (*g*) +. (*h*) −. (*i*) −. (*j*) +.

10. (*a*) $+0{\cdot}8660$. (*b*) $-0{\cdot}6583$. (*c*) $-13{\cdot}30$. (*d*) $-1{\cdot}62$. (*e*) $-0{\cdot}4098$. (*f*) $+3{\cdot}458$. (*g*) $+0{\cdot}2748$. (*h*) $-0{\cdot}0317$. (*i*) $-1{\cdot}432$. (*j*) $-0{\cdot}3421$.

11. (*a*) $\bar{1}{\cdot}9727$. (*b*) $\bar{1}{\cdot}9722$. (*c*) 0·3789. (*d*) 0·5412. (*e*) 0·1252. (*f*) 0·1342.

12. (*a*) 18° 19′. (*b*) 64° 47′. (*c*) 114° 49′. (*d*) 19° 32′. (*e*) 18° 5′. (*f*) 108° 19′.

Exercise LXI

1. (*a*) $A=65°\ 14'$; $B=24°\ 46'$; $b=15{\cdot}67$.
(*b*) $B=55°\ 45'$; $b=619{\cdot}1$; $c=749$.
(*c*) $A=42°\ 45'$; $a=3{\cdot}132$; $b=3{\cdot}388$.
(*d*) $A=68°\ 12'$; $B=21°\ 48'$; $c=10{\cdot}77$.
(*e*) $A=21°\ 10'$; $c=1010$; $b=942$.
(*f*) $B=24°\ 30'$; $A=65°\ 30'$; $a=3{\cdot}153$.
(*g*) $B=46°\ 30'$; $b=16{\cdot}25$; $a=15{\cdot}42$.
(*h*) $B=53°\ 14'$; $C=36°\ 46'$; $c=82{\cdot}44$.
(*i*) $B=33°\ 31'$; $A=56°\ 29'$; $c=4264$.
(*j*) $B=52°\ 37'$; $A=37°\ 23'$; $a=27{\cdot}31$.

2. 21° 5′.
3. 9·158 ft.
4. (*a*) 49·71 ml. (*b*) 97·58 ml.
5. 27·54 ml.
6. (*a*) 49 ft. 9 in. (*b*) 41 ft. 8 in.
7. 18,369 ft.
8. 2100 ft.
9. (*a*) 38·97 ml. (*b*) 1207·5.
10. 244½° T.
11. 6° 41·1′.
12. 9° 57·5′.
13. 114·5 ml.
14. 6·72.
15. 9·328 ml.
16. 0·1850.
17. 639 yd.
18. 12° 19′.
19. 51° 50′.
20. 203·6 m.p.h.

Printed in the United States
By Bookmasters